FOREWORD

Site characterisation and evaluation are important elements for determining the site suitability and long-term safety of geological repositories for radioactive waste disposal. They also provide vital information for the design of the underground facility and the engineered barrier system that will contain the waste. In this context, the NEA Radioactive Waste Management Committe (RWMC) established the Co-ordinating Group on Site Evaluation and Design of Experiments for Radioactive Waste Disposal (SEDE) to address the many issues related to site characterisation and evaluation.

One of the initiatives of the SEDE Co-ordinating Group has been to organise a Workshop on the investigation of possible long-term geological changes at potential repository sites. These investigations are designed to analyse the present characteristics of the site, their differences in the geological past and their likeliness to change in the futures. When considered in site characterisation and evaluation programmes, this information indicates the geological stability of the repository area, and can serve as the base for developing scenarios and boundary conditions needed in analysing the long-term safety at the repository.

The Workshop was held on 19-21 September 1994 at the OECD headquarters in Paris, immediately before the Fifth Meeting of the SEDE Co-ordinating Group. The objectives were:

- to review experience in characterising long-term geological changes in national programmes;
- to develop an understanding of the coupling and the relationships between geological processes;
- to discuss methods of hypothesis testing in the characterisation of long-term geological changes; and
- to develop ideas for planning and carrying out measurements of these long-term changes.

This Workshop acknowledged the importance of gathering information on long-term geological change not only for assessing the safety of deep repositories but also for siting and designing such facilities. The necessity of a multidisciplinary approach was also clearly stressed.

These proceedings reproduce the papers presented at the Workshop and include an executive summary. The opinions, conclusions and recommendations expressed are those of the authors only, and do not necessarily represent the views of any OECD Member country or international organisation. They are published on the responsibility of the Secretary-General of the OECD.

TABLE OF CONTENTS

(*) Paper not available.
(**) Extended Abstract only.

SESSION III
Chairman: J.-F. Aranyossy (France)

SESSION IV
Chairman: T. Äikäs (Finland)

EXECUTIVE SUMMARY

Introduction

On 19th-21st September 1994, in Paris, the NEA Co-ordinating Group on Site Evaluation and Design of Experiments for Radioactive Waste Disposal (SEDE) conducted a Workshop aimed at addressing technical and strategic issues related to the characterisation of long-term geological changes being considered for the evaluation of potential repository sites for the disposal of radioactive waste. The Workshop was organised by the OECD Nuclear Energy Agency assisted by a Programme Committee consisting of T. Äikäs (Finland), M. Thury (Switzerland), M. Raynal (France), T. McEwen (United Kingdom) and A. Lappin (United States). E. Patera, C. Pescatore and P. Lalieux have assured, alternatively, the secretariat of the SEDE Co-ordinating Group. This executive Summary has been drafted on the basis of the wrap-up discussion which closed the Workshop. The aims of this discussion were to summarise the experience to date and make, to the extend possible, recommendations for the direction of future investigations into characterising long-term geological changes.

This Workshop focused on the experience of the national programmes carrying out such investigations. It was requested that presentations included a review of which long-term changes were being investigated, how they influence each other and what the consequences for repository siting, safety assessment and public acceptance might be.

Presentations

In his Introductory Statement A. Hooper (United Kingdom), Chairman of the SEDE Co-ordinating Group, welcomed the Workshop participants and reminded that while we would be unwise to predict the future, we should pass the test of reasonableness when indicating likely trends. In order to achieve this, a holistic approach is probably required, where information from independant sources can be integrated.

In his presentation, D.T. Hoxie (United States) described the potentially disruptive features, events and processes which could alter hydrogeological conditions in the unsaturated zone of the Yucca Mountain Tuffs throughout a nominal repository system lifetime of 10 000 years or more. Evidence collected to date indicates that preferential flow pathways are likely to be present within the tuffs and that some of them may be transmitting water into the unsaturated zone under present-day climatic conditions. An extensive programme of paleoenvironmental and paleohydrological studies has been undertaken to assess past climatic changes in the region, the likelihood of future changes and their consequences on the hydrogeological system. Groundwater upwelling from the saturated zone was analysed in depth, and eventually discounted as a potential disruptive event.

M. Wallner (Germany) showed an example of probabilistic structural computations for assessing consequences of a temperature decrease (glacial age) on the integrity of the geosphere barrier for a repository located within a salt dome. Even under very conservative assumptions, thermally induced fractures will only develop as far as about 100 meters deep, and therefore will not endanger the barrier function of the salt. Sensitivity analysis were used to determine parameters with significant influence on fracture development at the top of the salt dome. Based on this example, the probabilistic modelling tool was considered as being efficient for evaluating consequences of uncertain future events especially because it allows uncertainty and sensitivity analysis and interrelation of input parameters.

In his presentation, H.-J. Herbert (Germany) analysed four typical long-term mineralogical changes in salt formations. For two processes, "Hartsalz" and "Carnallitit" dissolution, a positive correlation was found between the results of large scale in situ experiments and the results of geochemical modelling. A good agreement between geological and mineralogical observation and geochemical calculations in the case of the formation of the gypsum cap rock on the top of the Zechstein salt formations, and the polyhalitization of anhydrite was also demonstrated. Potential contribution of these studies to the correct evaluation of the brines' origin and of the hazards which brine inflows may pose for the safety of a underground repository in rock salt was also discussed.

C. del Olmo Alonso (Spain) synthetised nine years of studies related to geodynamic and climate research in the Iberic peninsula aimed at developing a geoprospective-surveying research methodology supporting safety assessment of radioactive waste facilities in deep geological formations within the framework of the Spanish geographical environment. Based on climatic evolution and environmental changes on the Iberian Peninsula and surroundings Mediterranean countries during the Quaternary, a paleoenvironmental reconstruction for a specific site (within the Tajo Valley) over the last 100 000 years has been carried out and is currently used as a basis for the construction of future evolution scenarios at the same paleosite for the next 100 000 years.

J.-F. Aranyossy (France) presented the geoprospective approach followed in the framework of the preliminary investigation phase carried out in four different locations in France. This approach would lead to the assessment of the role of geodynamic processes responsible for the geological evolution of each study area. Then, this past geological evolution would be extrapolated to the future at different time scale relevant to the waste disposal issue. Two examples illustrated the above-mentioned approach: a) the development of a new Würn-type glaciation and its consequences on the groundwater circulation, and b) the occurrence of a paroxysmic geodynamic event and its erosonial consequences.

M. Thury (Switzerland) detailed the geological evolution studies and neotectonic field investigations which have lead to an assessment of the effects of long-term geological changes on a potential repository in the crystalline basement of Northern Switzerland. These potential effects have been quantified and conservative maximum values have been estimated for the next million years. The consequences of the expected long-term changes on groundwater flow in the crystalline basement and on repository siting and safety were also analysed. It was concluded that negative effects on a repository can be avoided by adequate siting and design.

After giving an introduction on the structure and geology of the Fennoscandian Shield, P. Vuorela (Finland) described the current prevailing continuous slow processes affecting the Finnish bedrock and the approaches utilised for evaluating long-term changes. Examples of conservative scenarios used in assessing the performance of a potential repository were also given. It was concluded that although it is not possible to predict the future behaviour of a site in a detailed manner, it is possible to constrain the long-term evolution scenarios needed in safety assessment by studying actual events that occurred in the past.

L.O. Ericsson (Sweden) explained the programme aimed at quantifying the consequences of earthquake, glaciation and land uplift for the safety of a final repository for spent nuclear fuel within the Swedish crystalline bedrock. The emphasis was put on the modeling of future hydrogeological conditions in the occurrence of glaciation. For that purpose, a 3D ice sheet model has been developed for Scandinavia and coupled to a subglacial groundwater flow model which provides boundary conditions for evaluations of long-term hydrogeological evolution at specific sites.

T. McEwen (United Kingdom), in place of L.M. King et al., detailed the construction of long-term evolution scenarios to assist in the evaluation of a hypothetical deep repository for spent-fuel located in Äspö (SKI's SITE-94 project). The so-called Central Scenario allows evaluation of the climate and consequent surface and subsurface environments at the site for the next c. 120 000 years. It provides a first indicator of the physical and hydrogeological conditions below and at the front of the advancing and retreating ice sheets. The numerous estimates and assumptions made during the scenario development were particularly stressed.

The programme undertaken by Nirex aimed at estimating possible future changes to the groundwater flow system at Sellafield in Cumbria was detailed by R. Chaplow (United Kingdom). Efforts are being made to develop models of the site evolution which can be tested by seeking to predict current observations on the basis of processes which have influenced the site in the past. Analysis of the past history of subsidence and uplift, the evolution of the fractures controlling groundwater flow, the Quaternary evolution of the area and paleohydrogeological studies are underway.

C.C. Davison (Canada) presented studies to assess the geological stability of plutonic rocks of the Canadian Shield to assist in siting a disposal facility. A first series of studies included current seismicity monitoring and the assessment of the occurrence probability of an earthquake large enough to affect the integrity of the geosphere surrounding a potential repository. In addition, the potential for future fracture propagation at a potential disposal site was assessed by developing an understanding of the fracture history of the rockmass at the site. For this reason, fracture mineral infillings and alteration assemblage were thoroughly studied. A practical example of the application of these methodologies was presented.

P. Flavelle (Canada) presented the regulatory position adopted by the AECB relative to long-term geological changes for the Canadian waste disposal programme. He proposed that bounding calculations on the maximum possible impact of such processes on repository safety should in first instance be made. If these calculations reveal negligible impact, then further consideration of dynamic geosphere and long-term geological changes is unnecessary.

Wrap-up Discussion

The wrap-up discussion was introduced by T. Äikäs (Finland) and summarised by the SEDE Chairman at the Fifth Meeting of the Co-ordinating Group which was held just after the Workshop.

According to the presentations made, it is clear that information concerning long-term geological evolution is needed for several different purposes:

- Disposal concept design (repository depth and dimensions, materials to be used, ...);
- Safety assessment (scenario development, degree of conservatism, ...);
- Confidence building (technical and public acceptance, ...); and
- Siting programme (site identification, ...)

Although the Workshop presentations were somewhat biased towards crystalline media (evolution of fracture system) and the assessment of glaciation sequences, several national organisations showed that they have well developed programmes and presented many new results and considerations that are important to future developments of site characterisation activities.

From a technical point of view, the main points made at the workshop are as follows:

- most sites are not in equilibrium;
- the consideration of long-term geological changes is viewed as important to safety;
- there is a trend towards utilising more and more geochemical and hydrochemical techniques especially in the framework of the hypothesis testing methodology; the use of isotope techniques (and more precisely, the comparison of information delivered by several isotopes studies) is clearly acknowledged as are the important improvements in isotopic analysis techniques in the recent years. Within this context, the importance of representative sampling is specially singled out;
- approaches are being developed to obtain a consistent picture on different spatial scales of the long-term geological evolution (detailed description of site evolution requires to be placed in a more global/regional context);
- there is recognition that past evolution has brought about present day characteristics and that we must show a good understanding of the past and present conditions to build confidence in the treatment of future evolution;
- there is recognition that geologists can contribute to the assessment of long-term evolution but that the longer time periods are, the less scientific basis there is to develop probabilities of occurrence.

The main outcomes of the Workshop which are of interest for the working programme of national organisations in the field of the assessment of long-term stability of the geosphere concern:

- the development of approaches to describe evolution that are fit for the purpose;
- the assessment of over-conservatism (which could have the effect to eliminate the consideration of the geosphere as a barrier);
- the consensus building at all informed levels;
- the confidence building in the use of isotopic analysis and paleohydrogeological methods in support of hypothesis testing;
- the appropriate coupling of processes in evolution modeling;
- the use of natural analogues (most considerations to date, however, have been limited only to permafrost and volcanism); and
- the discussion of time frames to be addressed (with the regulatory organisations).

Potentially Disruptive Hydrologic Features, Events and Processes at the Yucca Mountain Site, Nevada

Dwight T. Hoxie
U.S. Geological Survey
Las Vegas, Nevada, USA

ABSTRACT

Yucca Mountain, Nevada, has been selected by the United States to be evaluated as a potential site for the development of a geologic repository for the disposal of spent nuclear fuel and high-level radioactive waste. If the site is determined to be suitable for repository development and construction is authorized, the repository at the Yucca Mountain site is planned to be constructed in unsaturated tuff at a depth of about 250 meters below land surface and at a distance of about 250 meters above the water table. The intent of locating a repository in a thick unsaturated-zone geohydrologic setting, such as occurs at Yucca Mountain under the arid to semi-arid climatic conditions that currently prevail in the region, is to provide a natural setting for the repository system in which little ground water will be available to contact emplaced waste or to transport radioactive material from the repository to the biosphere. In principle, an unsaturated-zone repository will be vulnerable to water entry from both above and below. Consequently, a major effort within the site-characterization program at the Yucca Mountain site is concerned with identifying and evaluating those features, events, and processes, such as increased net infiltration or water-table rise, whose presence or future occurrence could introduce water into a potential repository at the site in quantities sufficient to compromise the waste-isolation capability of the repository system.

INTRODUCTION

Burial in underground repositories sited in suitable geologic settings is considered generally throughout the international community to be the preferred method for the permanent disposal of radioactive waste [1]. The primary function of the geologic setting is to provide a repository host rock and a stable natural setting that, in combination with an appropriately designed engineered-barrier system, will impede the movement of radioactive material from the repository to the biosphere. Siting repositories in thick unsaturated zones in arid regions has been proposed to offer a number of advantages not only for long-term waste isolation but also for repository construction, operation, and waste retrievability [2,3]. The United States (U.S.) currently is proceeding with the concept of an unsaturated-zone repository for the disposal of spent nuclear fuel and high-level radioactive waste. In 1987 the U.S. Congress directed the U.S. Department of Energy (DOE) to evaluate a site in the arid southwestern U.S. at Yucca Mountain, Nevada, to determine the suitability of this site for development of a mined geologic repository. If the site is determined to be suitable and a license is granted by the U.S. Nuclear Regulatory Commission for repository construction and operation, the repository at the Yucca Mountain site is planned to be excavated in unsaturated tuff at a depth of about 250 m below land surface and at a distance of about 250 m above the water table and would have a design capacity for 70,000 metric tons of heavy-metal waste.

Waste isolation in an unsaturated-zone repository under arid climatic conditions would be achieved principally by providing a natural setting in which little ground water would be available to contact emplaced waste and to transport radioactive material to the biosphere. This waste-isolation concept is based on the premise that moving ground water is the primary means by which radionuclides are likely to be transported from a geologic repository to the biosphere and the subsidiary hypotheses that in an unsaturated-zone setting (1) little of the water held in storage by capillarity and adsorption within an unsaturated host medium will be available to contact emplaced waste and (2) under sustained arid climatic conditions, negligible quantities of ground water will be moving through the repository either to contact emplaced waste or to transport radionuclides.

Reliance on this waste-isolation concept implies that a potential failure mode for an unsaturated-zone repository is the presence or occurrence of features, events and processes that could introduce large quantities of water into the repository or its surrounding host rock. A major effort within the site-characterization program that the DOE is conducting at the Yucca Mountain site is being devoted to determining present and past geohydrologic conditions at the site and to assessing possible and expected future changes in these conditions that could impact waste isolation during a nominal 10,000-year or longer repository lifetime. This paper describes the geohydrologic setting of the Yucca Mountain site and presents a summary report of progress achieved to date in identifying and evaluating such features, events, and processes whose presence or occurrence during the next 10,000 years could alter hydrologic conditions within the unsaturated zone sufficiently to compromise the waste-isolation capability of a potential, unsaturated-zone repository at the site.

This report was prepared with support provided by the Yucca Mountain Site Characterization Project Office of the U.S. Department of Energy under Interagency Agreement DE-AI08-92NV10874.

GEOHYDROLOGIC SETTING OF THE YUCCA MOUNTAIN SITE

Yucca Mountain is located about 160 km northwest of Las Vegas, Nevada, (Figure 1) in the southern part of the Great Basin. The Great Basin is a subregion within the Basin and Range physiographic province, a broad region of crustal extension in the western and southwestern U.S. and northern Mexico consisting of tilted, fault-block mountain ranges and deep alluvium-filled intermontane basins. The interior of the Great Basin, in particular, is characterized by linear, generally north-trending, fault-bounded mountain ranges separated by broad (20 to 30 km) alluvial valleys and basins. Although crustal extension continues at present within the Great Basin, the rate of extension has slowed from a maximum of about 20 to 30 mm/yr in the interval 10 to 15 ma to about 10 mm/yr since 5 ma, and the locus of active tectonism has migrated westward to the Owens Valley, Death Valley, and Long Valley Caldera regions along the western margin of the Great Basin near the border between Nevada and California [4]. This evidence suggests the concurrent decline of both the rate and magnitude of tectonism in the Yucca Mountain area since the occurrence of the volcanic and structural events that created Yucca Mountain in middle-to-late Miocene time.

Yucca Mountain evolved from a thick depositional apron of predominantly rhyolitic ash-flow and ash-fall tuffs and minor intercalated bedded tuff that were erupted episodically during middle Miocene time (10 to 14 ma) from a caldera complex whose eroded remnants adjoin Yucca Mountain on the north. Faulting and tilting during or immediately succeeding late-stage caldera collapse, together with subsequent erosion, has produced the plexus of north- to northwest-trending, parallel to subparallel ridges and intervening canyons and valleys that constitute the present-day physiography of Yucca Mountain. West-dipping normal faults bound the major north-south ridges and demarcate the western boundaries of individual, usually east-tilted structural blocks within the Yucca Mountain complex.

The potential repository at Yucca Mountain is proposed to be constructed beneath a prominent ridge and east-tilted structural block within the central part of the Yucca Mountain complex (Figure 1). At the site of the potential repository, the ridge, designated the Yucca Crest in Figure 1, attains an altitude of about 1,500 m above sea level and stands more than 300 m above the valley floors to the west and east. The structural block in which the potential repository would be constructed is bounded on the west by a steep erosional escarpment that developed along the trace of a major west-dipping normal fault, on the east by north-trending zones of imbricate normal faults, and on the north by northwest-trending strike-slip faults. Strata within the block dip variably from 5° to 10° to the east.

The stratigraphy of Yucca Mountain consists of unconsolidated surficial materials (alluvial, colluvial, and eolian deposits and thin veneers of residuum) overlying a layered sequence of variably welded, altered, and fractured tuffs of middle-Miocene age. The unconsolidated surficial materials range in thickness from zero on bedrock outcrops to several tens of meters or more in the canyon and valley floors. The Tertiary volcanics section attains an aggregate thickness of 2 km or more and lies unconformably on Paleozoic carbonate rocks.

Based largely on the degrees of welding and alteration of the tuffs, the rock-stratigraphic units within the unsaturated zone and uppermost part of the saturated zone at the Yucca Mountain site have been grouped into informally named hydrogeologic units [5]. In descending order from land surface, with the ranges of thickness of the units indicated in parentheses, these hydrogeologic units are: the Tiva Canyon welded unit (0 to 150 m); the Paintbrush nonwelded unit (20 to 100 m); the Topopah Spring welded unit (290 to 360 m); the Calico Hills nonwelded unit (100 to 400 m); and the Crater Flat undifferentiated unit. The Crater Flat unit designates the sequence of moderately to densely welded tuffs of unspecified thickness that underlie the Calico Hills unit and constitute much of the upper part of the saturated zone beneath the potential repository site. The Tiva Canyon and Topopah Spring welded units are composed of moderately to densely welded, devitrified, fractured (10 to 40 fractures/m^3) ash-flow tuff that is characterized by low rock-matrix porosity (~0.1) and low rock-matrix hydraulic conductivity (~10^{-11} m/s). The Paintbrush nonwelded unit consists of unfractured (~1 fracture/m^3), nonwelded to partially welded ash-fall tuff and bedded tuff of relatively high matrix porosity (~0.4) and high hydraulic conductivity (~10^{-7} m/s). The Calico Hills unit consists of mainly unfractured (2 to 3 fractures/m^3), nonwelded to partially welded ash-flow and ash-fall tuff and bedded tuff of moderate matrix porosity (~0.3). The initially glassy tuffs within the Calico Hills unit, however, have been altered extensively to zeolities and other minerals progressively from south to north in the site area; and the matrix hydraulic conductivity of the unit ranges from about 10^{-7} m/s, where the unit is predominantly vitric, to about 10^{-10} m/s, where it has been zeolitized. The tuffs composing the Crater Flat unit are generally fractured (8 to 25 fractures/m^3) and of moderately low matrix porosity (~0.2) and hydraulic conductivity (~10^{-9} m/s).

The potential repository at Yucca Mountain is proposed to be constructed in densely welded tuff in the lower part of the Topopah Spring welded unit. The units above the potential repository horizon, specifically the Paintbrush nonwelded and Tiva Canyon welded units, together with the unconsolidated surficial materials where present, are hydrologically important because these units control the spatial distribution and flux of ground water that enters the unsaturated zone as net infiltration and percolates downward through the unsaturated zone toward the water table. The Calico Hills nonwelded and Crater Flat undifferentiated units below the potential repository constitute the principal hydrologic and geochemical barriers for water-borne transport of radionuclides from the potential repository to the biosphere. The contrasting hydrologic properties between and within these units are expected to produce distributions of water content and ground-water flux within the unsaturated zone that are likely to be highly variable in both space and time.

The configuration of the water table at the Yucca Mountain site is marked by two distinctive and as yet unexplained features. The water table is virtually flat and stands at an altitude of about 730 m above sea level beneath and to the east and south of the potential repository location. Immediately to the north of the potential repository, however, the water table rises steeply in altitude by about 300 m over a lateral distance of about 3 km. The region of small hydraulic gradient may indicate the presence of high-transmissivity materials, low ground-water flux, or a combination of these two factors. The region of large hydraulic gradient is more enigmatic, although several hypotheses have been advanced to explain its presence [6]. The ground-water flow system within the saturated zone beneath Yucca Mountain is of importance because the most likely route by which radioactive material released from a potential Yucca Mountain repository may reach the biosphere is by vertically downward transport by ground water moving through the unsaturated zone to the water table and subsequently by lateral transport and dispersal in the saturated zone.

Recharge to the ground-water flow system by water percolating downward through the unsaturated zone at Yucca Mountain appears to be virtually negligible. Present-day average annual precipitation at Yucca Mountain ranges from 150 to 170 mm, depending principally on altitude [7] .

Average potential evapotranspiration at Yucca Mountain, however, is estimated to be about 1,500 to 1,700 mm [8], which suggests that most of the water incident on Yucca Mountain as precipitation is returned to the atmosphere by evapotranspiration with only a small residual remaining to enter the unsaturated zone as net infiltration. The hydraulic conductivity values for the tuffaceous rocks composing Yucca Mountain imply that vertical ground-water flux through the unsaturated zone must be less than about 1 mm/yr in order to sustain unsaturated conditions within the rock matrix. Preliminary hydrologic modeling [9] that represents ground-water flow as steady-state Darcian flow through an unsaturated rock matrix further indicates that average ground-water fluxes through the unsaturated zone must be in the range from 0.01 to 0.1 mm/yr in order to match the saturation profiles observed in deep boreholes at the site. Studies of water content in the shallow unsaturated zone at the site, however, indicate present-day near-surface fluxes that range from 0.02 mm/yr to as much as 13.4 mm/yr depending on spatial location [10]. These data, together with recent observations of perched-water occurrences in deep unsaturated-zone boreholes, indicate that both the spatial and temporal distributions of water content and ground-water flux are likely to be highly variable within the unsaturated zone at Yucca Mountain. The resultant complexly three-dimensional state of the unsaturated-zone hydrologic system is the product of both present and past geologic and hydrologic processes whose continuance or change will govern future states of the system and their consequent implications for waste isolation in the unsaturated zone at Yucca Mountain.

POTENTIALLY DISRUPTIVE FEATURES, EVENTS, AND PROCESSES

The intent of locating a repository in the unsaturated zone at Yucca Mountain is to provide a geohydrologic setting in which little water will be available to contact emplaced waste and to transport radionuclides from the repository to the biosphere. Although available evidence indicates that such a setting exists within the unsaturated zone at Yucca Mountain under present-day conditions, the critical issue for long-term waste disposal at the site concerns the expectation that the extant geohydrologic setting will be sustained throughout a nominal repository-system lifetime of 10,000 years or more. In this context, the following three sets of features, events, and processes have been identified whose presence or occurrence at Yucca Mountain could alter hydrologic conditions in the unsaturated zone sufficiently to warrant evaluation of their potential future impacts on waste isolation at the site: (1) the presence of preferential flow pathways that could provide conduits for localized ground-water flow into or through the unsaturated zone; (2) the occurrence of future climatic change that could increase net infiltration into the unsaturated zone; and (3) the occurrence of tectonically or climatically induced water-table rise into or near an unaturated-zone repository at the site. Progress achieved to date in evaluating these possible features and occurrences is summarized below.

Preferential Flow Pathways

The geologic setting, consisting of faulted, tilted structural blocks composed of layered sequences of fractured welded tuffs alternating with relatively unfractured nonwelded tuffs, endows the geohydrologic framework at Yucca Mountain with pronounced heterogeneity and anisotropy. Heterogeneity results from the juxtaposition of hydrogeologic units of highly contrasting hydrologic properties at stratigraphic boundaries and fault contacts and from the presence of discontinuities across fractures and faults. Anisotropy is manifest in the tilted, layered stratigraphy and the common alignment of fractures within fracture networks. Within this framework the distribution and movement of water within the unsaturated zone not only is likely to be complexly three dimensional but may be governed locally by nonequilibrium conditions and processes as well [11]. In particular, models of ground-water flow that are based on Darcian concepts of capillarity-driven flow in unsaturated porous-media may not be adequate to account for the inherent heterogeneity. Consequently, an alternative conceptual model has been proposed [12] that considers transient, gravity-driven flow in highly localized preferential flow pathways to be a major, if not dominant, mode of water movement into and through the unsaturated zone at Yucca Mountain. Candidates for such preferential flow pathways include the fracture systems within the welded tuffs, fractured zones associated with the faults that bound and transect the structural blocks, and high-saturation zones within relatively permeable nonwelded tuff units.

Evidence for the presence of preferential flow pathways in the unsaturated zone at Yucca Mountain derives from several sources. The occurrences of perched-water bodies [13], high-saturation zones [14], and anomalous concentrations of ^{3}H [15] and ^{36}Cl [16] that have been observed in boreholes imply that, even under present-day arid climatic conditions, water is being channeled into the unsaturated zone and is able to descend to depths ranging from several tens to hundreds of meters. Presently dry fractures throughout the unsaturated zone contain coatings and fillings of calcite that bears a pedogenic isotopic signature and apparently was precipitated from water that had passed through the soil zone and had continued downward in the fractures [17]. Pneumatic testing in boreholes [18] shows that the fractured tuffs are characterized by high rock-mass bulk permeabilities, which are apparently fracture dominated and indicate that the fractures form extensive integrated and hydraulically connected flow networks. Finally, observations of active localized flow systems and pathways in settings similar to that of Yucca Mountain [19], including spring discharge from apparently fault-controlled perched-water systems and ground-water flow from fractures intersected by tunnels excavated in otherwise unsaturated tuffaceous rocks, provide indirect evidence for the likely presence of pathways for localized ground-water flow within the unsaturated zone at Yucca Mountain.

The implications for waste isolation at a potential Yucca Mountain repository posed by the presence of preferential flow pathways in the unsaturated zone remain to be addressed fully. Such pathways may be viewed as being beneficial if they were to divert downward moving ground water away from the potential repository and its emplaced waste, but would be detrimental if they were to channel flow towards the repository. Because of the limited quantities of water that are available to enter the unsaturated zone under present-day climatic conditions, the presence of potential preferential flow pathways above an unsaturated-zone repository may be viewed as benign unless and until they are activated by an influx of water, for example, from increased net infiltration in response to future climatic change. However, as a consequence of waste-generated heat release following repository development, these flow pathways may figure prominently in redistributing in situ moisture and establishing a thermal-hydrologic flow regime in the unsaturated zone whose effects may dominate and overwhelm any naturally occurring change within the hydrologic system for periods up to or, perhaps, exceeding 1,000 years after repository closure [20].

Future Climatic Change

The thick unsaturated zone at Yucca Mountain owes its presence in part to the arid to semi-arid climatic conditions that prevail in the southern Great Basin. This climatic regime provides little effective moisture (defined as precipitation minus evapotranspiration) to enter the unsaturated zone as net infiltration. Consequently, to the extent that the maintenance of essentially present-day conditions in the unsaturated zone at Yucca Mountain is to be relied on to isolate radioactive waste at the site, the long-term stability of the extant climatic regime emerges as an issue of major concern. However, because climatic change is not strictly a geologic disruptive process, only a brief summary of an approach for evaluating climatic change at Yucca Mountain will be discussed here.

Of primary concern is future climatic change that would increase net infiltration and introduce water into the unsaturated zone in sufficient quantities and rates to reach an unsaturated-zone repository, contribute to waste-package degradation and failure, and transport radionuclides to the biosphere. The central issue of future climatic change, therefore, is threefold: (1) How much water entering at land surface and moving through the unsaturated zone, either as Darcian flow in the rock mass or as episodic or sustained flow in preferential pathways, is too much water? (2) What degree of climatic change, both in magnitude and duration vis-a-vis the present-day climatic regime, would be required to introduce such quantities of water into the unsaturated zone at Yucca Mountain? (3)

What is the likelihood that climatic change of such degree will occur in the Yucca Mountain region during a nominal repository lifetime of 10,000 years?

Performance-assessment evaluations that are being conducted for a potential Yucca Mountain repository will provide a basis for addressing the first question. The second question is difficult to address because the transfer function relating climatic variables to water entry into the unsaturated zone is not straightforward and remains to be determined for the Yucca Mountain site. Once the issues associated with the first two questions are resolved, however, the third question can be addressed, in principle, by climate-modeling studies to estimate the likelihood of occurrence of those global and regional conditions and forcing functions that could combine to generate climatic change of the required degree. This inverse approach for evaluating future climatic change would not necessarily entail the development of detailed climate models but, instead, would focus on examining the plausibility of climatic change based on the specific consequences that such change would have for waste isolation at Yucca Mountain.

An intensive program of paleoenvironmental studies has been undertaken to support reconstructions of climatic and hydrologic conditions and their interrelationship during the Holocene and late Pleistocene Epochs in the Yucca Mountain region. Knowledge of past climatic conditions and change is essential for calibrating and testing models intended to predict future climatic change and to establish bounds on the extremes of naturally induced climatic variation that may be expected in the future. Evaluation of the evidence for changing hydrologic conditions will contribute toward developing the functional relation between climatic change and hydrologic-system response. There is, of course, no assurance that future climates will mimic past climates, especially in view of the uncertain effects of increased concentrations of greenhouse gases in the atmosphere as a result of human activity. Consequently, knowledge of past climatic conditions and change provides a guide to, but not necessarily an analog for, future climatic conditions and change.

Water-Table Rise

By virtue of being located in the unsaturated zone, the potential repository at Yucca Mountain will be subject potentially not only to ground-water entry from above but also to ground-water entry from below. Ascending water would be manifest as local or regional water-table rise that could occur either quasi-statically, for example, in response to climatically induced increased recharge, or dynamically as a consequence of tectonic or hydrothermal processes. The possibility of dynamically induced water-table rise to or above the altitude of the potential repository at Yucca Mountain has emerged as a major source of controversy in assessing the suitability of the site for potential repository development.

In 1989, J.S. Szymanski, who was then with the Department of Energy, released a draft report [21] in which he suggested that upwelling ground water from the saturated zone has breached the surface in the Yucca Mountain area repeatedly in geologically recent times. The principal evidence cited by Szymanski to support this hypothesis is the presence at and near Yucca Mountain of (1) near-surface and subsurface fractures and faults filled with calcite and subordinate opaline silica veins, (2) breccias composed of angular bedrock fragments cemented by calcite and opaline silica, and (3) surface-parallel deposits of calcium carbonate within the soil zone. Szymanski proposed that these deposits were produced by mineral-laden water ascending from the saturated zone and discharging at the surface. Szymanski further proposed that tectonic stress release ("seismic pumping") and (or) hydrothermal convection provided the mechanisms to force water vertically upward from the saturated zone. These mechanisms, however, confront serious obstacles in that to be viable they must be capable of forcing appreciable quantities of water upwards from the water table through several hundred meters of unsaturated rock. Furthermore, the most likely source of water with high calcium-carbonate content is the Paleozoic carbonate rocks that underlie the thick Tertiary volcanic section at Yucca Mountain. In order for water from the carbonate rocks to reach the surface, however, the water would have to ascend through 2 km or more of saturated and unsaturated volcanic rocks.

Because of the negative implications for waste isolation at a potential Yucca Mountain repository posed by this so-called "Szymanski hypothesis," the DOE conducted an intensive review of Szymanski's data and conclusions and supported numerous studies to determine the origin of the calcite-silica deposits cited by Szymanski as evidence. Controversy arose because the DOE investigators maintained that the deposits in question originated from near-surface processes [24] analogous to those responsible for calcrete and caliche formation that is a common occurrence in desert environments [22]. The controversy culminated in a review of the contending hypotheses by a panel convened for the purpose by the U.S. National Research Council, which issued a report [23] that strongly endorsed the interpretation of a pedogenic origin for the near-surface calcite-silica deposits.

The evidence against upwelling ground water as the source of the near-surface calcite-silica deposits derives principally from the morphology of the deposits and their distinctive isotopic composition. These deposits bear little resemblance to the travertine, tufa, and siliceous sinter ("geyserite") deposits that typically form around warm and hot springs by the degassing of CO_2-charged waters ascending from depth. In contrast, the deposits in question consist of fine-grained, micritic calcite; low-temperature opal-CT; numerous root casts, ooids, and pellets; and inclusions of 20 percent or more of detrital material [24], which is consistent with a low-temperature origin by evaporative precipitation of minerals that are dissolved and mobilized locally in meteoric water moving through the soil zone. To ascribe the mineralogy and morphology of these deposits to the surface discharge of deep-seated ground water would seem to entail chemical and physical processes heretofore unobserved in the Yucca Mountain region or elsewhere.

The isotopic composition of the near-surface calcite-silica deposits provides independent evidence for a shallow, pedogenic origin of these deposits. The stable isotopes ^{18}O and ^{16}O (as well as ^{13}C and ^{12}C) are partitioned between calcite and ground water during calcite precipitation and are diagnostic of the composition and temperature of the water from which a particular calcite precipitated. Unless altered by chemical interaction with the rocks through which the water passes, the relative abundance of the isotopes ^{18}O and ^{16}O in ground water generally reflects the isotopic composition of the meteoric water entering the ground-water flow system as recharge and, thus, the environmental conditions under which recharge occurred. The isotope pairs ^{87}Sr and ^{86}Sr and ^{234}U and ^{238}U are not partitioned during calcite precipitation; consequently, the relative abundances of these isotopes are direct indicators of the source of the precipitated calcite.

Measurement of stable-isotope concentrations in water samples obtained from 21 boreholes that penetrate the saturated zone beneath Yucca Mountain yield a mean $\delta^{18}O$ value of -13.52 ± 0.4 ‰ (SMOW) [25], and present-day precipitation at Yucca Mountain has a mean $\delta^{18}O$ value of about -9.4 ‰ (SMOW) [25]. The disparity between the isotopic composition of Yucca Mountain ground water and contemporary precipitation at Yucca Mountain has been interpreted to reflect differences between the climatic regime that prevailed when recharge to the ground-water flow system beneath Yucca Mountain occurred and the climatic regime that currently prevails in the southern Great Basin [25]. These $\delta^{18}O$ data indicate that Yucca Mountain ground water and Yucca Mountain meteoric water represent two isotopically distinct water bodies whose isotopic imprint should be imparted to minerals precipitated from these waters. In particular, if the near-surface calcite-silica deposits at Yucca Mountain were formed by ground water ascending from the saturated zone beneath Yucca Mountain, calcites within these deposits should reflect the $\delta^{18}O$ composition of that water. Water temperatures at the water table beneath Yucca mountain average about 30 °C [26]. If calcite were to precipitate at these temperatures under equilibrium conditions from Yucca Mountain ground water with mean $\delta^{18}O = -13.5$ ‰, the resulting calcite should have $\delta^{18}O \leq 12$ ‰ [24]. Measurements of $\delta^{18}O$ on calcite samples from the near-surface calcite-silica deposits at Yucca Mountain, however, yield $\delta^{18}O$ values in the range from 19 to 22 ‰ [24]. In order for Yucca Mountain ground water to precipitate calcites with $\delta^{18}O$ values in the range observed for the near-surface calcite samples, calcite precipitation would have to have occurred at temperatures less than 5 °C [24], which is not consistent with the hypothesis that these calcites derived from warm upwelling water from the saturated zone. On the other hand, if the near-surface calcites were to have precipitated at a soil temperature of about 10 °C from water with $\delta^{18}O$ = -9.4 ‰, corresponding to the mean composition of present-day rainfall and snowmelt at Yucca Mountain, the resulting calcite would have a $\delta^{18}O$ value of about 21 ‰ [24], which is consistent with the observed $\delta^{18}O$ values for the near-surface calcites. Although the isotopic composition of calcites formed in the soil zone by carbonate dissolution in and precipitation from infiltrating meteoric water will depend complexly on many factors [27], the $\delta^{18}O$ evidence for the near-surface calcites strongly favor a hypothesis of local pedogenic origin as opposed to direct precipitation from ground water ascending from the saturated zone.

Available strontium isotope data for ground water in the saturated zone beneath Yucca Mountain and for the near-surface calcite-silica deposits also are incompatible with the hypothesis that these deposits derived from water that ascended from the saturated zone. Because of the accuracy with which the isotope ratio $^{87}Sr/^{86}Sr$ can be measured (±0.00005 or better) and because these isotopes are not fractionated during calcite precipitation, the $^{87}Sr/^{86}Sr$ ratio provides a reliable indicator of the water from which a particular calcite precipitated [24]. Measured values of $^{87}Sr/^{86}Sr$ for ground water in the saturated zone beneath Yucca Mountain range from 0.7093 to 0.7113. The near-surface calcite-silica deposits, however, tend to be more radiogenic: $^{87}Sr/^{86}Sr$ values for these calcites range from 0.7116 to 0.7127 [28]. The hypothesis of calcite-silica deposition from upwelling ground water from the saturated zone offers no ready explanation for the apparent ^{87}Sr enrichment in the near-surface deposits. A more reasonable interpretation is that these deposits are of pedogenic origin and that the observed $^{87}Sr/^{86}Sr$ values resulted from the interaction between locally infiltrating meteoric water and surficial material enriched in ^{87}Sr (or depleted in ^{86}Sr) relative to Yucca Mountain ground water [24].

Uranium isotope data also fail to support the hypothesis that the near-surface calcite-silica deposits at Yucca Mountain derived from ground water from the saturated zone beneath Yucca Mountain. Like strontium, the uranium isotopes do not fractionate during calcite precipitation and, thus, are indicators of the source water from which the calcite precipitated. Once uranium is immobilized, for example, by inclusion in calcite, the subsequent radioactive decay will tend towards secular equilibrium, and the $^{234}U/^{238}U$ activity ratio will approach 1.0. In particular, calcites precipitating from ground water will have the same $^{234}U/^{238}U$ activity ratio as the parental ground water. Calculated initial $^{234}U/^{238}U$ activity ratios on calcite samples from the near-surface calcite-silica deposits at Yucca Mountain are all less than 1.5; whereas water from the saturated zone beneath Yucca Mountain has $^{234}U/^{238}U$ activity ratios that range from 5.00 to 6.94 [24]. These data indicate that ground water in the saturated zone is enriched in ^{234}U and could not be the source water for the near-surface calcite deposits. $^{234}U/^{238}U$ activity ratios determined for soils in the Yucca Mountain area generally are less than 1.4 with only a single value as high as 2.0 [29]. These soil data suggest that the observed initial $^{234}U/^{238}U$ activities in the near-surface calcite-silica deposits derived from a pedogenic source. In any case, the $^{234}U/^{238}U$ activity data do not support the hypothesis that the near-surface calcite-silica deposits originated from saturated-zone ground water.

Although the geochemical data cited above do not support the hypothesis of geologically recent water-table rise sufficient to breach the surface at Yucca Mountain, the potential for possible future water-table rise to the level of the potential repository in the unsaturated zone at Yucca Mountain remains to be addressed. In this regard, it is entirely appropriate to seek evidence of former water-table altitudes to guide assessments of the likelihood for and magnitude of changes in water-table altitude that may occur in the future. Two lines of geochemical evidence, which are discussed briefly below, indicate that although the water table beneath Yucca Mountain has been at higher levels in the past, it probably has not stood higher than about 60 to 100 m above its present configuration with respect to the structural and stratigraphic setting of Yucca Mountain since this setting was created by faulting and tilting of the Yucca Mountain block during the interval 11.6 to 12.8 ma.

One line of geochemical evidence for past water-table altitudes derives from the alteration history of the tuffs at Yucca Mountain [30]. Core samples from boreholes indicate that the stratigraphic section at Yucca Mountain is transected laterally by a persistent zone that is about 10 m thick and records a progressive downward transition from unaltered tuff above to altered tuff below. In particular, below the transition zone, initially glassy nonwelded tuffs have been extensively altered, predominantly to the zeolite mineral clinoptilolite and subordinately to mordenite, clays, and other minerals. The alteration appears to have been a diagenetic process in which the original glass pyroclasts were replaced by zeolites under conditions of abundant water and at ambient temperatures. Based on the consequent implication that the zeolitization occurred below the water table at the time of alteration, the present-day configuration of the alteration transition zone embeds a record of former maximum water-table altitudes with respect to present-day structure and stratigraphy at Yucca Mountain. A preliminary reconstruction of this record [30] suggests that maximum water-level altitudes may have been attained at Yucca Mountain in the period 11.6 to 12.8 ma and, since that time, have not been greater than about 60 m above present-day levels for periods sufficiently long ($\sim 10^4$ yr) to induce zeolitization of the vitric tuffs.

A second line of evidence is provided by analyses of secondary calcite that pervasively, but not ubiquitously, occurs as fracture and vug fillings within both the unsaturated and saturated zones at Yucca Mountain. The data clearly indicate that calcites in the saturated zone are morphologically, chemically, and isotopically distinct from the calcites present in the unsaturated zone [17,31,32], a reflection of both differing precipitation environments and differing source materials. Consequently, the distribution of saturated-zone and unsaturated-zone calcites can be used, in principle, to map the location and altitude of former high-stands of the water table.

Strontium isotope data provide the best example to date for the use of calcite analyses to infer possible former water-table altitudes at Yucca Mountain. $^{87}Sr/^{86}Sr$ values determined [28] for 16 calcite samples obtained from the unsaturated zone at altitudes more than 100 m above the present-day water table range from 0.7115 to 0.7127. This range of values coincides with the range observed for pedogenic calcite at Yucca Mountain and implies that secondary calcite in the unsaturated zone derived from a pedogenic source. In contrast, $^{87}Sr/^{86}Sr$ values for 19 calcite samples from the saturated zone at depths from 250 to 1,275 m below the water table range from 0.7087 to 0.7098, which clearly distinguishes these calcites from those deposited in the unsaturated zone. However, four calcite samples collected from the unsaturated zone within 100 m of the present-day water table yielded $^{87}Sr/^{86}Sr$ values that ranged from 0.7108 to 0.7112. These $^{87}Sr/^{86}Sr$ values are significantly less than the values obtained from samples higher in the unsaturated zone and indicate the presence of a non-pedogenic contribution to strontium in these samples. The most likely candidate source for the anomalous $^{87}Sr/^{86}Sr$ values is ground water from the saturated zone; consequently these data may indicate that the water table at the borehole site from which the samples were taken has been higher by as much as 100 m above the present level at this site. Data from other locations are needed, however, in order to determine whether this apparent anomaly represents a trend over the Yucca Mountain area or is an isolated occurrence.

Evidence for changing water-table altitudes near Yucca Mountain are indicated from examinations of three paleospring deposits that are located about 20 km southwest of the Yucca Mountain site. Uranium-series dating of calcites from these deposits indicate that, although currently inactive, water discharged at these sites at 18 ± 1, 30 ± 3, 45 ± 4, and > 70 ka [33]. Initial $^{234}U/^{238}U$ activity ratios for these calcites range from 2.8 to 3.8 and suggest that water from the ground-water flow system in the volcanic rocks underlying Yucca Mountain supplied the water that formed these deposits. At the present time, the water table is interpolated to be about 80 to 115 m below the spring deposits, thus implying that when these springs were active the water table locally stood at least 80 to 115 m higher than present-day levels. The magnitude of possible changes of water-table altitude beneath Yucca Mountain that may have occurred during times of ground-water discharge from these paleosprings is unknown but, perhaps, could be estimated from ground-water modeling studies.

The geochemically based inferences of former water-table altitudes at Yucca Mountain are indirect but are corroborated in part by data indicating that the potentiometric surface in the regional ground-water flow system has undergone progressive decline since at least mid-Pleistocene time [34]. The data are based on uranium-series dating of vein calcites associated with sites of former spring discharge from the carbonate aquifer. Depending on the degree of hydraulic connection between the two flow systems, lowering of the potentiometric surface in the regional carbonate-rock aquifer would be expected to induce water-table decline in the saturated-zone ground-water flow system overlying the carbonate-rock aquifer beneath Yucca Mountain. These data, based additionally on occurrences of abandoned spring-discharge areas well above the present potentiometric surface [35], imply regional potentiometric-surface declines ranging from tens to possibly hundreds of meters during the Quaternary Period. Possible explanations for such apparent declines include tectonic lowering of the discharge areas, especially in the Death Valley region; tectonic uplift of the abandoned spring sites; water-table lowering due to increased aridity or erosion; or some combination of these processes [34]. The apparent decline of the regional potentiometric surface is consistent with and may have contributed to secular decline of water-table altitudes beneath Yucca Mountain.

SUMMARY AND CONCLUSIONS

The United States has selected Yucca Mountain, Nevada, as a potential site for the Nation's first geologic repository for the disposal of spent nuclear fuel and high-level radioactive waste. The potential repository at Yucca Mountain is planned to be excavated in unsaturated tuffaceous rocks at a depth below land-surface of about 250 m and at a distance above the local water table of about 250 m. Locating a repository in a thick unsaturated-zone geohydrologic setting under arid to semi-arid climatic conditions, such as occur at the Yucca Mountain site, is intended to provide a a natural setting for the potential repository in which little water will be available either to contact emplaced waste or to transport radioactive material from the repository to the biosphere. The concept of siting geologic repositories in natural settings that will restrict the flow of ground water into and out of the repository system is common to virtually all radioactive-waste-disposal programs; however, reliance on a thick unsaturated zone in an arid environment to accomplish this function is unique to the U.S. program.

By virtue of being located in the unsaturated zone, however, a repository at Yucca Mountain will be subject, in principle, to water entry from both above and below. The entry of water in sufficient quantities to promote waste-package degradation and failure and subsequent radionuclide mobilization and transport has been recognized to constitute a possible failure mode for a potential Yucca Mountain repository system. Consequently, in characterizing the Yucca Mountain site to determine the suitability of the site for potential repository development, considerable effort has been expended to evaluate those features, events, and processes that could cause water to be introduced into the repository from either above or below and lead to possible repository-system failure.

In order for water to invade an unsaturated-zone repository from above, two conditions must be met. First, there must be a source of water and, second, there must be pathways present by which to convey water into and through the unsaturated zone to the repository. In general, net infiltration from rainfall and snowmelt constitutes the source of water that enters the unsaturated zone; however under present-day climatic conditions of low precipitation and high evapotranspiration, little water is available to infiltrate the unsaturated zone at Yucca Mountain. Consequently, future climatic change that would lead to increased net infiltration at Yucca Mountain emerges as an event of considerable importance to potential future waste isolation at the Yucca Mountain site. In order to address the issue of possible future climatic change, an extensive program of paleoenvironmental and paleohydrologic studies has been undertaken to assess not only the occurrence, magnitude, and duration of past climatic change in the Yucca Mountain region but also the consequences of such change on the hydrologic systems. Using past climatic change as a basis, climate modeling studies will be conducted to assess the likelihood of future climatic change at Yucca Mountain.

Increased net infiltration alone, however, need not impair the waste-isolation capability of an unsaturated-zone repository if the time for the effects of increased infiltration to reach the repository is long compared to the design lifetime of the repository system. Because the rocks exposed over much of the surface of Yucca Mountain as well as those in which the potential repository would be located are densely welded tuffs of low saturated hydraulic conductivity ($\leq 10^{-11}$ m/s), the welded-tuff rock matrix is unlikely to sustain ground-water fluxes sufficient to deliver appreciable quantities of water from land surface to the potential repository in times that are of concern. However, Yucca Mountain is a faulted, titled block of alternating welded and nonwelded tuffs of highly contrasting hydrologic properties, and the welded tuffs at Yucca Mountain tend to be highly fractured (20 to 40 fractures/m^3). Consequently, ground-water flow into and through the unsaturated zone is likely to be complexly three dimensional and may tend to be channeled into localized preferential flow pathways consisting of interconnected fracture systems, fault zones, or nonwelded tuff units of relatively high saturated hydraulic conductivity ($\sim 10^{-7}$ m/s). Evidence collected to date indicate not only that such pathways are likely to be present at Yucca Mountain but also that some of these pathways may be active under present-day climatic conditions and be capable of transmitting water from tens to hundreds of meters into the unsaturated zone. The issue to be resolved with respect to water entry from above the potential repository, therefore, entails evaluating the likelihood of future climatic change that, coupled with the presence of preferential flow pathways both above and below the repository, could lead to unacceptable releases of radioactive material to the biosphere.

The possibility of water entry from below the potential repository, in particular as a result of dynamic, tectonically or hydrothermally induced, water-table rise, has been a subject of controversy at the Yucca Mountain site. Occurrences of near-surface calcite-silica deposits in the Yucca Mountain area, consisting of veins within fault zones, cementing material in breccias, and slope-parallel deposits, have been interpreted by some to have been produced by recurrent episodes of ground water upwelling from the saturated zone beneath Yucca Mountain. An alternative interpretation, however, considers these deposits to have resulted from near-surface processes of dissolution and precipitation of calcium carbonate from meteoric water infiltrating and percolating through the soil zone. Support for the latter interpretation is provided by the widespread occurrence in desert soils of deposits of calcium carbonate as caliche and calcrete. Additional evidence for a pedogenic origin of the near-surface calcite-silica deposits at Yucca Mountain, however, derives from their geochemistry, specifically their oxygen, strontium, and uranium isotopic compositions, which virtually precludes modern ground water from the saturated zone beneath Yucca Mountain as the source water for these deposits. These geochemical data, however, are fully consistent with a near-surface, pedogenic origin for these deposits. These data, together with Ockham's razor, which enjoins one to accept the simplest hypothesis consistent with the data, support the hypothesis of near-surface origin for the calcite-silica deposits and obviate the need to devise mechanisms, such as seismic pumping or hydrothermal convection, to propel water from the saturated zone through hundreds of meters of overlying unsaturated rock to breach the surface at Yucca Mountain.

To discount upwelling ground water from the saturated zone as the source of the near-surface calcite-silica deposits, however, is not to say that the water table at Yucca Mountain has not stood higher in the past or that it will not stand higher than its present level in the future. Incomplete and somewhat indirect evidence to date, based on an analysis of water-induced mineral alteration in the tuffs beneath Yucca Mountain and limited strontium isotope data, indicates that the water table beneath Yucca Mountain has been higher in the past but is not likely to have been higher than about 60 to 100 m above its present level since the interval 11.6 to 12.8 ma when major tectonic deformation of the Yucca Mountain block occurred. This inference is supported by additional evidence that potentiometric levels within the Yucca Mountain region have undergone secular decline throughout the Quaternary Period or longer.

The most credible naturally occurring event by which water may be caused to enter a potential repository in the unsaturated zone at Yucca Mountain would seem to be future climatic change that would appreciably increase net infiltration and recharge in the Yucca Mountain area. The magnitude and duration of climatic change that would be required to introduce sufficient water into the unsaturated zone to cause a potential Yucca Mountain repository system to fail is at present unknown but probably would require appreciably increased rates of precipitation and decreased rates of evapotranspiration relative to present-day rates. The likelihood of occurrence of such a climatic event during the design life of the potential repository system is difficult to assess, not only because predictions of climatic change are inherently uncertain but also because the effects on future climates due to greenhouse-gas emissions and other human activities are largely unknown. The uncertainty attaching to predictions of future climatic change and the response of the unsaturated-zone hydrologic system to such change probably are the principal contributors to the overall uncertainty in evaluating the long-term performance of a potential repository in the unsaturated zone at Yucca Mountain in response to natural occurrences.

REFERENCES

1 Witherspoon, P.A., ed.: "Geological Problems in Radioactive Waste Isolation, A World Wide Review," Proceedings of Workshop W3B, 28th International Geological Congress, Washington D.C., 1989, 233 p.

2 Winograd, I.J.: "Radioactive Waste Disposal in Thick Unsaturated Zones," Science, 212, 4502 (1981), 1457-1464.

3 Roseboom, E.H., Jr.: "Disposal of High-Level Nuclear Waste above the Water Table in Arid Regions," U.S. Geological Survey Circular 903, 1983, 21 p.

4 Wernicke, B., Axen, G.J., Snow, J.K.: "Basin and Range Extensional Tectonics at the Latitude of Las Vegas, Nevada," Geological Society of America Bulletin, 100 (1988), 1738-1757.

5 Montazer, P., and Wilson, W.E.,: "Conceptual Model of Flow in the Unsaturated Zone, Yucca Mountain, Nevada," U.S. Geological Survey Water-Resources Investigations Report 84-4345, 1984, 55 p.

6 Fridrich, C.J., Dudley, W.W., Jr., and Stuckless, J.S.: "Hydrogeologic Analysis of the Saturated-Zone Ground-water System, under Yucca Mountain, Nevada," Journal of Hydrology, 154 (1994), 133-168.

7 Hevesi, J.A., Flint, A.L., and Istok, J.D.: "Precipitation Estimation in Mountainous Terrain Using Multivariate Geostatistics, Part II: Isohyetal Maps," Journal of Applied Meteorology, 31 (1992), 677-688.

8 Kohler, M.A., Nordenson, T.J., and Baker, R.D.: "Evaporation Maps, for the United States," U.S. Weather Service/U.S. Department of Commerce, 1959.

9 Bodvarsson, G., Chen, G., and Wittwer, C.: "Preliminary Analysis of Three-Dimensional Moisture Flow Within Yucca Mountain, Nevada," Proceedings of the Fifth Annual International High-Level Radioactive Waste Conference, 1994, 2038-2047.

10 Flint, A.L., and Flint L.E.: "Spatial Distribution of Potential Near Surface Moisture Flux at Yucca Mountain," Proceedings of the Fifth Annual International Conference on High-Level Radioactive Waste Management, 1994, 2352-2358.

11 Nitao, J.J., and Buscheck, T.A.: "Infiltration of a Liquid Front in an Unsaturated, Fractured Porous Medium," Water Resources Research, 27, 8 (1991), 2099-2112.

12 Gauthier, J.H., Wilson, M.L., Peters, R.R., Dudley, A.L., and Skinner, L.H.: "Total System Performance Assessment Code (TOSPAC) Volume 2: User's Guide," Sandia National Laboratories Report SAND85-0004, 1992, 204 p.

13 Luckey, R.R.: "Update on Fluid in USW UZ-14," Presentation to Nuclear Waste Technical Review Board, 19-20 October 1993, Las Vegas, NV.

14 Flint, A.: "Characterization of Infiltration," Presentation to Nuclear Waste Technical Review Board, 11-12 December 1989, Denver, CO.

15 Yang, I.C.: "Flow and Transport through Unsaturated Rock -- Data from Two Test Holes, Yucca Mountain, Nevada," Proceedings of the Third Annual International High-Level Radioactive Waste Management Conference, 1992, 732-737.

16 Fabryka-Martin, J.T., Wightman, S.J., Murphy, W.J., Wickham, M.P., Caffee, M.W., Nimz, G.J., Southon, J.R., and Sharma, P.: "Distribution of Chlorine-36 in the Unsaturated Zone at Yucca Mountain: An Indicator of Fast Transport Paths," Proceedings of FOCUS '93: Site Characterization and Model Validation, 1993, 58-68.

17 Vaniman, D.T.: "Calcite Deposits in Fractures at Yucca Mountain, Nevada," Proceedings of the Fourth Annual International Conference on High-Level Radioactive Waste Management, 1993, 1935-1939.

18 LeCain, G.D., and Walker, J.N.: "Results of Air-Permeability Testing in a Vertical Borehole at Yucca Mountain, Nevada," Proceedings of Fifth Annual International High-Level Radioactive Waste Management Conference, 1994, 2782-2788.

19 Thordarson, W.: "Perched Water in Zeolitized Bedded Tuff, Ranier Mesa and Vicinity," U.S. Geological Survey Open-File Report TEI-862, 1965, 90 p.

20 Buscheck, T.A., and Nitao, J.J.: "The Analysis of Repository-Heat-Driven Hydrothermal Flow at Yucca Mountain," Proceedings of the Fourth Annual International High-Level Radioactive Waste Management Conference, 1993, 847-867.

21 Szymanski, J.S.: "Conceptual Considerations of the Yucca Mountain Groundwater System with Special Emphasis on the Adequacy of this System to Accommodate a High-Level Nuclear Repository," Unpublished Manuscript, 1989.

22 Machette, M.N.: "Calcic Soils of the Southwestern United States," Geological Society of America Special Paper 203, 1985, 1-21.

23 National Research Council: "Ground Water at Yucca Mountain Nevada, How High Can It Rise?", Final Report of the Panel on Coupled Hydrologic/Tectonic/Hydrothermal Systems at Yucca Mountain, 1992, 231 p.

24 Stuckless, J.S., Peterman, Z.E., Forester, R.M., Whelan, J.F., Vaniman, D.T., Marshall, B.D., and Taylor, E.M.: "Characterization of Fault-Filling Deposits in the Vicinity of Yucca Mountain, Nevada," Proceedings of the Waste Management '92 Conference, 1992, 929-935.

25 Benson, L., and Klieforth, H.: "Stable Isotopes in Precipitation and Ground Water in the Yucca Mountain Region, Southern Nevada," American Geophysical Union Geophysical Monograph 55, 1989, 41-59.

26 Sass, J.H., Lackenbruch, A.H., Dudley, W.W., Jr., Priest, S.S., and Monroe, R.J.: "Temperature, Thermal Conductivity, and Heat Flow near Yucca Mountain, Nevada: Some Tectonic and Hydrologic Implications," U.S. Geological Survey Open-File Report 87-649, 1988, 118 p.

27 McConnaughy, T.A., Whelan, J.F., Wickland, K.P., and Moscati, R.J.: "Isotopic Studies of Yucca Mountain Soil Fluids and Carbonate Pedogenesis," Proceedings of Fifth Annual International Conference on High-Level Radioactive Waste Management, 1994, 2584-2589.

28 Marshall, B.D., Peterman, Z.E., and Stuckless, J.S.: "Strontium Isotopic Evidence for a Higher Water Table at Yucca Mountain," Proceedings of the Fourth Annual International Conference on High-Level Radioactive Waste Management, 1993, 1948-1952.

29 Rosholt, J.N., Bush, C.A., Carr, W.J., Hoover, D.L., Swadley, W.C., and Dooley, J.R., Jr.: "Uranium-Trend Dating of Quaternary Deposits in the Nevada Test Site Area, Nevada and California," U.S. Geological Survey Open-File Report 85-540, 1985, 72 p.

30 Levy, S.S.: "Mineralogic Alteration History and Paleohydrology at Yucca Mountain, Nevada," Proceedings of the Third Annual International Conference on High-Level Radioactive Waste Management, 1992, 477-485.

31 Vaniman, D.T., and Whelan, J.F.: "Inferences of Paleoenvironment from Petrographic, Chemical, and Stable-Isotope Studies of Calcretes and Fracture Calcites," Proceedings of the Fifth Annual International Conference on High-Level Radioactive Waste Management, 1994, 2730-2737.

32 Whelan, J.F., Vaniman, D.T., Stuckless, J.S., and Moscati, R.J.: "Paleoclimatic and Paleohydrologic Records from Secondary Calcite: Yucca Mountain, Nevada," Proceedings of the Fifth Annual International Conference on High-Level Radioactive Waste Management, 1994, 2738-2745.

33 Paces, J. B., Taylor, E. M., and Bush, C.: "Late Quaternary History and Uranium Isotopic Compositions of Ground Water Discarge Deposits, Crater Flat, Nevada," Proceedings of the Fifth Annual International Conference on High-Level Radioactive Waste Management, 1994, 1573-1579.

34 Winograd, I.J., and Szabo, B.J.: "Water-Table Decline in the South-Central Great Basin During the Quaternary: Implications for Toxic Waste Disposal" in "Geological and Hydrologic Investigations of a Potential Nuclear Waste Disposal Site at Yucca Mountain, Southern Nevada," M.D. Carr and J.C. Yount, ed., U.S. Geological Survey Bulletin 1790, 1988, 147-152.

35 Winograd, I.J., and Doty, G.C.: "Paleohydrology of the Southern Great Basin with Special Reference to Water Table Fluctuations Beneath the Nevada Test Site during the Late(?) Pleistocene," U.S. Geological Survey Open-File Report 80-569, 1980, 91 p.

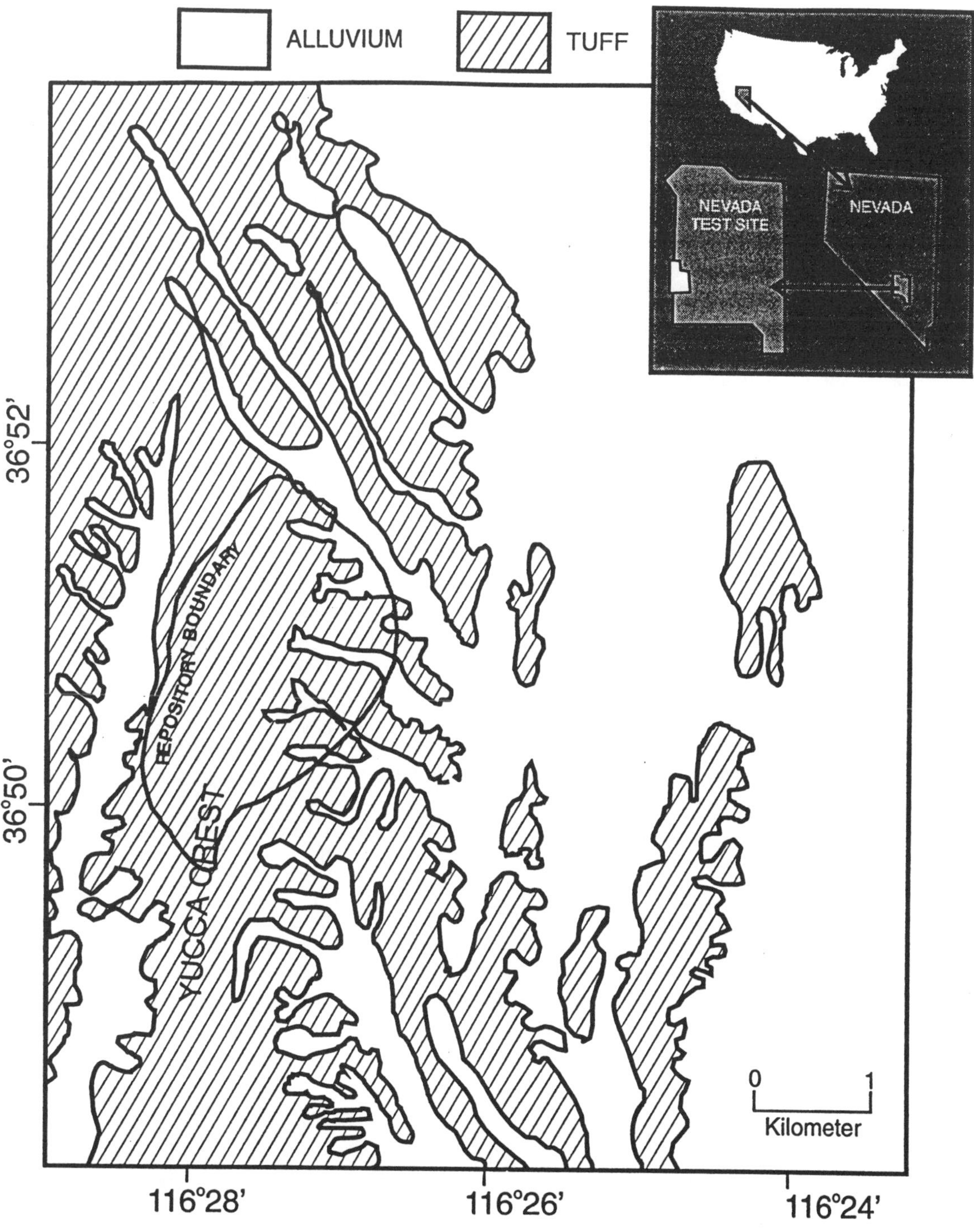

Figure 1. Map showing location of potential repository at Yucca Mountain, Nevada.

Modeling the Consequences of Future Geological Processes by Means of Probabilistic Methods

Manfred Wallner and Ralf Eickemeier

Federal Institute for Geosciences and Natural Resources
Hannover, Germany

Abstract: Safety assessment for a nuclear waste repository requires studying of risks regarding both long-term stability and serviceability of the repository, and long-term integrity of the geological barrier. Future geological processes have to be considered with respect to long-time performance of the overall disposal system. Assessment of structural safety is based on model computations. The purpose of the paper is to explain probabilistic modeling of structural response. As an example, probabilistic structural computations have been performed for an assumed glacial impact on the stability of a repository. The occurrence of tensile stresses at the top of the salt dome is regarded as characteristic results concerning the integrity of the salt barrier above the repository. The sensitivity of the results is discussed with regard to the variability of geotechnical input parameters.

1 Introduction

The salient objectives of the disposal of radioactive wastes in deep geological formations are [1]:

- to isolate radioactive wastes from the biosphere over long time scales without giving the responsibility to future generations to maintain the integrity of the disposal system, or imposing upon them significant constraints due to the existence of the repository, and

- to ensure long-term radiological safety, that is to protect humans and the environment against inadmissible radiation in accordance with agreed radiation protection principles.

These objectives imply that future risks shall not damage the containment given by the overall disposal system. Assessment of risks will ask for:

1. scenarios,
2. probability of occurrence of scenarios, and
3. consequences of these scenarios.

The overall disposal system consists of various components, such as the waste form, the repository, the backfill, engineered barriers and the host rock. The isolation of the radioactive waste is based on the multibarrier concept. However, for a repository in an intact salt formation the salt barrier may act as an encapsulated system, and therefore may provide the ultimate barrier. Even if one does not rest on the only functioning of the geological barrier, the most important scenario for a repository in rock salt is the loss of integrity of the salt barrier. Safety of the overall disposal system has to be demonstrated by predictive models considering relevant risk scenarios. Due to the long time-scale aspects, future events and processes also have to be taken into account.

The development of relevant scenarios [2] requires screening of all involved events and processes with respect to:

- probability of occurrence,
- intensity of consequence, and
- interrelation of scenarios.

The proper development of relevant scenarios is the most important step of geomechanical modeling. The method presented in the paper will deal with the consequence analysis for a given scenario. The performance of the overall disposal system is deterministically defined. However, the systems parameters whose values are uncertain are studied in a sensitivity analysis using probabilistic features.

2 Geotechnical design concept

The common principle of demonstrating safety in the field of civil engineering is to check that standardized maximum loading on a structure does not exceed permissible stresses. In most cases, this procedure cannot be applied to geotechnical problems. On the contrary, the fundamental idea of geotechnical engineering is to review the real circumstances of the geotechnical structure with criticism, and to recognize possible risk scenarios. According to safety reasons, it has to be demonstrated that all impacts on the structure are in a sufficient distance with respect to crucial risks and therefore do not exceed a safe state. Subsequently, geotechnical investigations do not consider standardized maximum loading and accordingly derived global values of safety but evaluate the site specific problem in a way that all relevant imposed loads (forces, displacements, temperature, etc.) and their impact on the underground structure is well understood, sufficiently considered and properly assessed [3].

The ultimate objective of the geotechnical investigations is to recognize risk conditions (approaching a limit state) and to take suitable actions to prevent those risks. For that reason, possible risks with respect to stability have mainly to be analyzed with regard to loss of bearing capacity. Possible risks with respect to the serviceability of the salt formation as a tight barrier has to be specially evaluated against loss of integrity. The assessment of structural safety will be based on model computations. Consequences of applied actions are also to be evaluated through model computations, if necessary stepwise, to minimize risks.
In this context, the following investigations and analyses are balanced elements of an overall assessment:

- geological exploration,
- geotechnical laboratory investigations,
- field experience,
- in situ measurements and
- model computations.

As to a safe design, geotechnical engineering aims at taking suitable actions (if required):

- to avoid risks, e.g. by changing the design,
- to improve the structural behavior e.g. through supporting systems,
- and/or to reduce risks to an acceptable level (residual risk).

3 Probabilistic structural analysis

Significant prerequisites for applying the above outlined geotechnical concept, e.g. for the design of a repository for radioactive wastes, are computational tools that give reliable results on the salt barrier's performance, especially for the long-time range.

Efficient geotechnical structural response computations will usually be carried out by means of finite element models. Generally, the variability of input parameters, like mechanical data or design variables, will be investigated in a parameter study. Although this procedure may yield in correct bounds for the geotechnical response with respect to the variation of the input parameter, it does not satisfy the demands of a comprehensive sensitivity study.

Under these circumstances it seems to be worthwhile to apply probabilistic methods and concepts for evaluating the significance of uncertainties on system performance. Respective tools, methods and concepts have been developed [4].

From a systematic point of view, "uncertainty" means "the lack of knowledge of the exact value or state" and may originate from:

- variability of genuine scattered geomechanical properties,
- insufficient testing of geotechnical data,
- modeling a complex geotechnical problem (simplification, homogenization, etc.),
- and variability of design features (e.g. maximal thermal loading, geometrical dimensions of the disposal fields).

To reduce the level of uncertainty, the following method presumes the application of a validated model and a verified computer code to run the computation. Moreover, human errors will not be taken into account because they are not a stochastic attribute.

Within a deterministic rock response model the consideration of randomly distributed input parameters will also lead to randomly distributed output results, e.g. displacements, strains and stresses. A stochastic computation based on a straight-forward Monte Carlo method very soon becomes inefficient for large FE-models, requiring a large number of random simulations. In practice, MC-method therefore is not often used for design purposes due to the vast amount of time-consuming simulations.

An alternative and efficient method for evaluating structural reliability and sensitivity is a probabilistic finite element method based on the so-called response surface technique [5] (s. App. A). The fundamental idea of this technique is that the system response y for any spatial coordinate and time is approximated by a function of the input parameters a_j. The system response surface Y is computed by means of a parameter study. The complex response surface is represented by a hyperplane which can be derived from a sufficient number of sample points. To avoid a complete factorial analysis for all input parameters to define the system response surface, the use of an orthogonal matrix related to an experimental design method [6] reduces the total number of system evaluations (s. App. B).

The sensitivity $Sens_i$ due to i-th input parameter is obtained by differentiating the response surface with respect to this parameter. The sensitivity therefore correspond to the tangent hyperplane of the true response function. The sensitivity describes how the system response changes if a single parameter is varied. Normally, the system response will be dependent on several parameters, which can be independent from each other and/or can be interrelated.

As an essential result from the computation of the sensitivities, the relative contribution of each parameters to the deviation of the system response can be determined. If we consider that the input parameters are random-variables whose probability density functions are given, the system response is also a random-variable and can be approximately calculated on the basis of stochastic features, namely expected value and standard deviation.

4 Simulation of glacial induced fractures

The ability of the presented method to perform a comprehensive sensitivity analysis shall be illustrated by means of an example simulating the impact of a temperature decrease during glacial age on the integrity of the salt barrier above the repository.

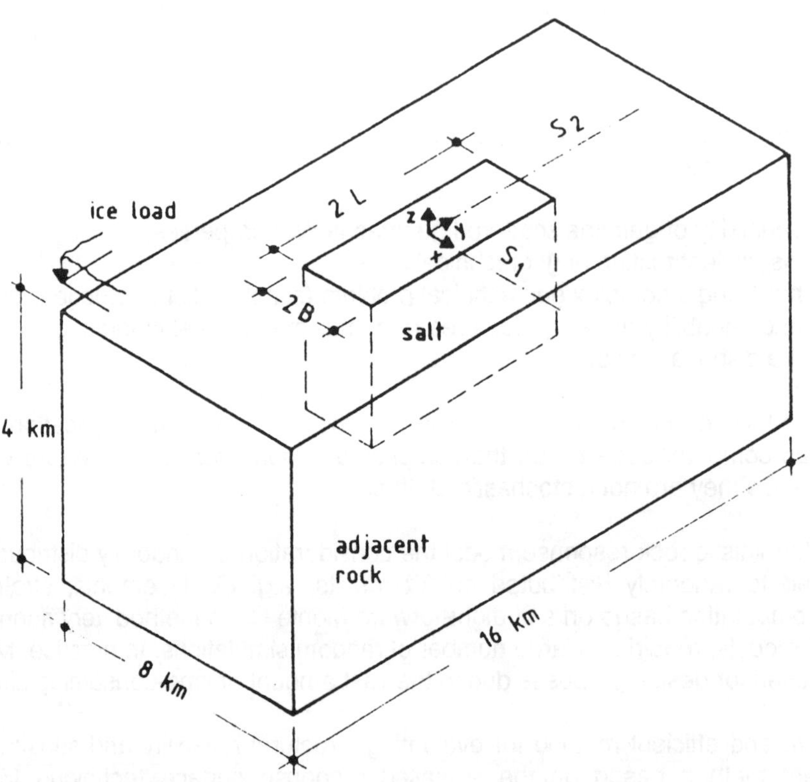

Figure 1. **Simplified geomechanical model (not to scale)**

Glacial induced fractures have been mapped at the top of several salt domes in Northern Germany [7]. As high level waste presents a potential hazard for very long times, it may be useful to study the consequences of a severe temperature decrease on the integrity of the salt barrier. Temperature induced fractures at the top of the salt dome may also result from temperature increase at the repository horizon due to decay heat. However, this process will occur over a relatively short time period after waste emplacement [8], so that both processes don't need to be superimposed but can be assessed separately.

Water-bearing strata in the overburden above the salt dome affect the consequences of the temperature decrease. Perma-frost will develop in the overburden which delays the temperature decrease at the top of the salt dome, and also damps down the total amount of temperature decrease [9]. However, for studying the fundamental effects, overlying strata are neglected in the model from a conservative viewpoint. A very simplified model applied for this purpose is shown in fig. 1.

The fundamental basis for a reliable assessment of the performance of the salt barrier is a consistent constitutive law for the mechanical behavior of rock salt. Rock salt shows a distinctive non-linear rate sensitive mechanical behavior that is particular dependent on stress and temperature. From an engineering point of view, however, the computation of long-term processes may be based on a simplified constitutive model which takes into account elastic behavior and steady state creep only. Failure deformation is not implicitly incorporated in the applied constitutive model. However, stresses are analyzed with respect to an appropriate yield-criterion.

The objective of the applied probabilistic modeling is to evaluate the intensity of thermally induced fractures as the consequence of:

- different amount of temperature decrease *T*,
- scattered data for the geotechnical features, namely stiffness E_S, creep A_S, and thermal expansion α of rock salt, as well as stiffness E_N of the adjacent rock, and
- variation of the geometrical dimensions *B* and *L* of the salt dome.

Geomechanical data:				
Young's Modules (rock salt)	:	E_S:	20,000 - 30,000 (Mpa)	
Creep parameter (rock salt)	:	A_S:	0.09 - 0.36	(d^{-1})
Thermal expansion coefficient (rock salt)	:	α :	$4.0 - 4.4 \cdot 10^{-5}$ (K^{-1})	
Young's Modules (adjacent rock)	:	E_N:	5,000 - 10,000	(Mpa)
Geometrical data:				
Width	:	*B*:	500 - 2,000	(m)
Length	:	*L*:	2,000 - 8,000	(m)
Temperature data:				
Temperature decrease	:	*T*:	15 - 20	(ºC)

Table 1. **Range of varied input parameter**

The current state of probabilistic modeling presumes that the rock response is a linear function of the above mentioned input parameters. However, interrelation between the geomechanical data is taken into account (s. App. B). The assumed range of the parameters is compiled in table 1.

Not as a restriction with regard to probabilistic modeling, rather than with respect to simplification, the geomechanical model used in the computations is based on the following assumptions:

- The top of the salt dome extends up to the surface.
- The simplified salt dome is assumed to be isotropic and homogeneous.
- The adjacent rock is assumed to behave elastically.
- The primary state of stress is considered to be isotropic and to increase with depth corresponding to the density of the rock mass.
- An increase of temperature with a gradient of 2 K per 100 meters depth is assumed.

The probabilistic modeling was done on a reduced three-dimensional model representing a quarter of the whole model. Along the symmetry planes, appropriate displacement boundary conditions have been adopted. The sensitivity study is composed of 16 computations based on the orthogonal matrix for this problem as presented in App. B. The computations were carried out with the finite element code ANSALT [10].

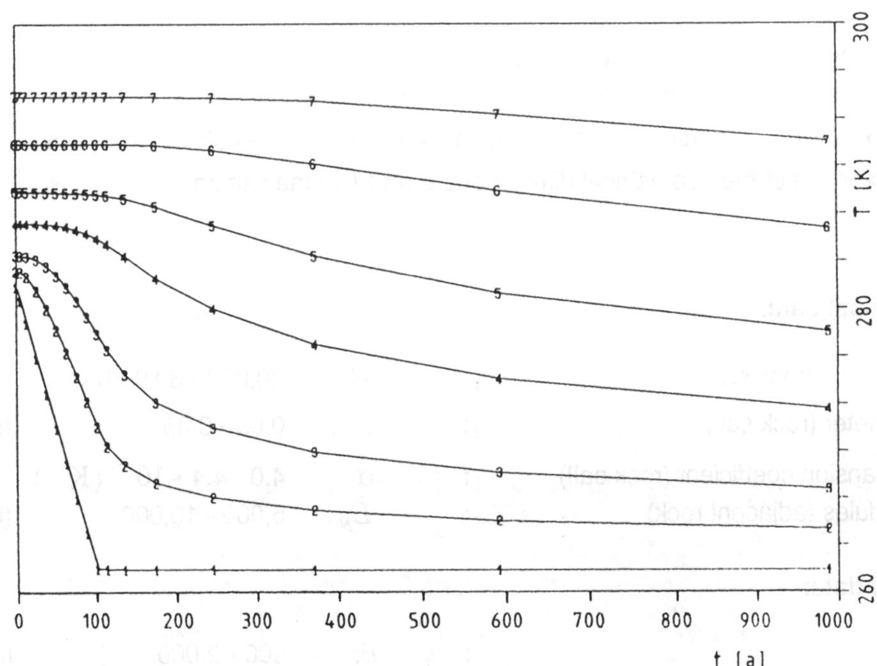

Figure 2. **Time history of temperature decrease in various depth of the salt dome**
(1) surface = 0 m (2) = 50 m (3) = 100 m (4) = 200 m
(5) = 300 m (6) = 450 m (7) = 600 m

5 Discussion of results

Considering previous experience, 3D-modeling for the salt integrity analysis turned out to be necessary to result in realistic and reliable representation of the involved processes. Although the applied model has been simplified, the fundamental load bearing behavior is sufficiently represented.

Characteristic results of the parameter study are shown in fig. 2 - 5 for the computed case no. 9 (s.App. B). Fig. 2 shows the time history of temperature in various depths of the salt dome as far as 600 meters below the surface. The correspondent time history for the horizontal stress at various depths of the salt dome is shown in fig. 3. Elastic behavior of the adjacent rock, and specially creep of rock salt enables the salt formation to reduce the total amount of tensile stress significantly. Nevertheless, tensile stresses do develop as far as almost 100 meters depth in this case. Unloading effects will even go deeper.

Stress profiles for the horizontal stress at the top of the salt dome (S1,S2) are plotted in fig. 4 for distinct times and in two orthogonal directions. As one can see, high stresses develop in the salt formation because of the higher thermal expansion coefficient of this layer compared to the adjacent rock. Those thermally induced stresses however have a maximum at about 100 years when the maximum temperature drop is considered to be reached and lateron decrease as a result of creep.

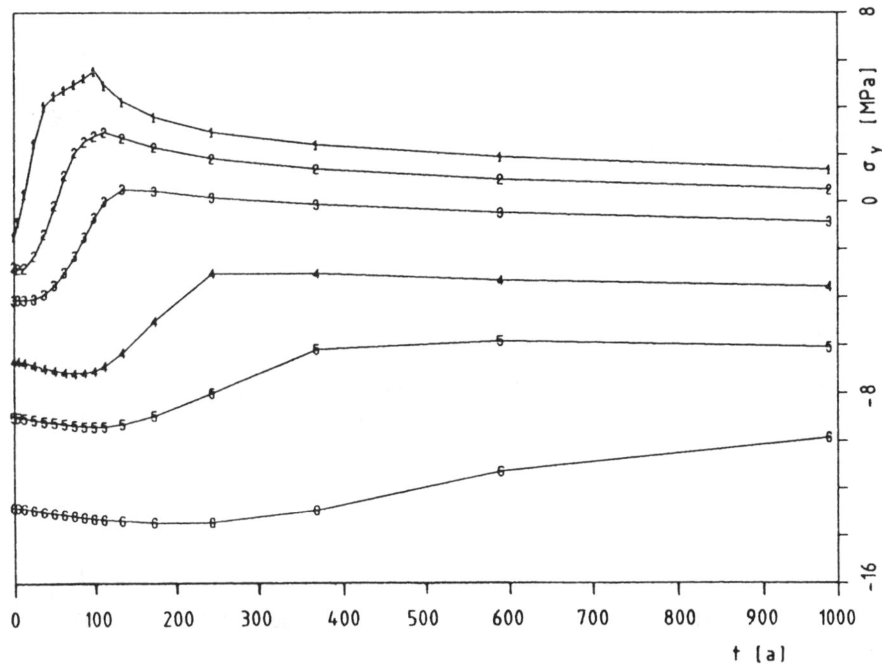

Figure 3. **Time history of horizontal stress σ_x in various depth of the salt dome**

(1) = 10 m (2) = 90 m (3) = 110 m

(4) = 290 m (5) = 320 m (6) = 580 m

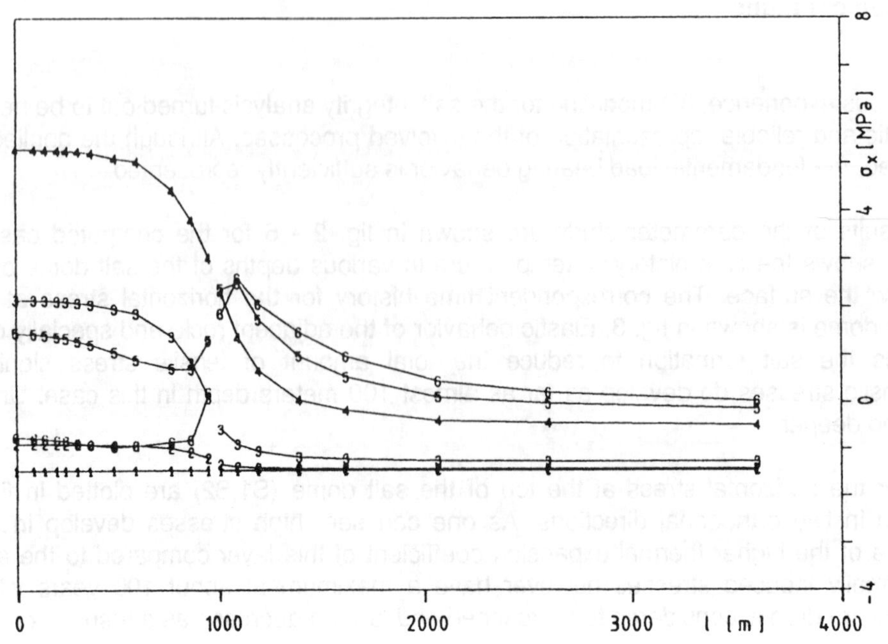

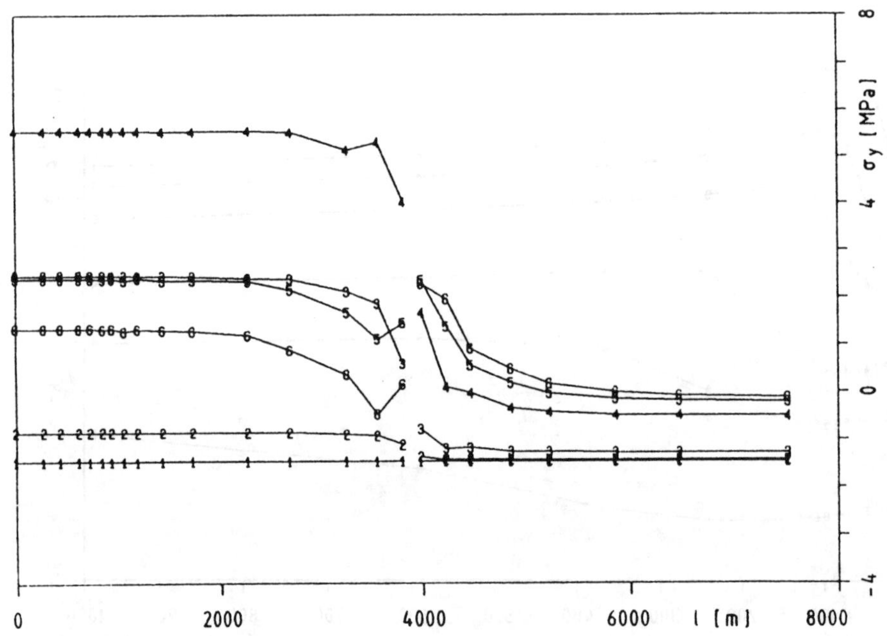

Figure 4. **Stress profiles at the top of the salt dome for distinct times**

(1) = 0 a	(2) = 5 a	(3) = 25 a
(4) = 100 a	(5) = 370 a	(6) = 1,000 a

The time history of surface subsidence is shown in fig. 5. Cooling of the rock salt causes contraction which results in a surface subsidence, and which reaches its maximum of 70 cm in the center of the salt dome after 1000 years.

The average contribution of the distinct parameters to the thermally induced stress σ_x is plotted in fig. 6. The time history plot is showing the influence of the single parameters and the influence of interrelated parameters, respectively. Obviously, the process is dominated by a direct influence of the creep capacity of rock salt, the stiffness of the adjacent rock mass, the width of the salt dome, and in the beginning by the amount of temperature drop. It is worthwhile to notice that the influence of interrelated mechanical data do not play an important role for the problem in question so that structural response can be approximated by a linear approach in this case.

Creep of rock salt has a negative sign influence on the stress indicating that high creep capacity will reduce the development of tensile stress due to temperature drop. On the other hand, stiffness of the adjacent rock, and width of the salt dome have a positive sign influence pointing at there significant influence to increase thermally induced stress with higher values of these input data.

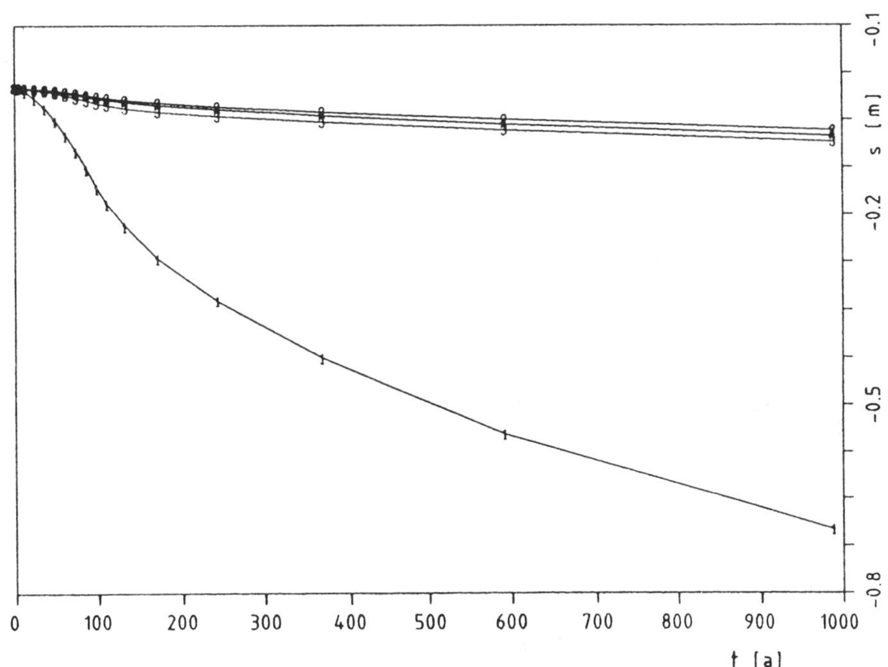

Figure 5. **Time history of surface subsidence**
(1) = center (2), (3), (4) = corners

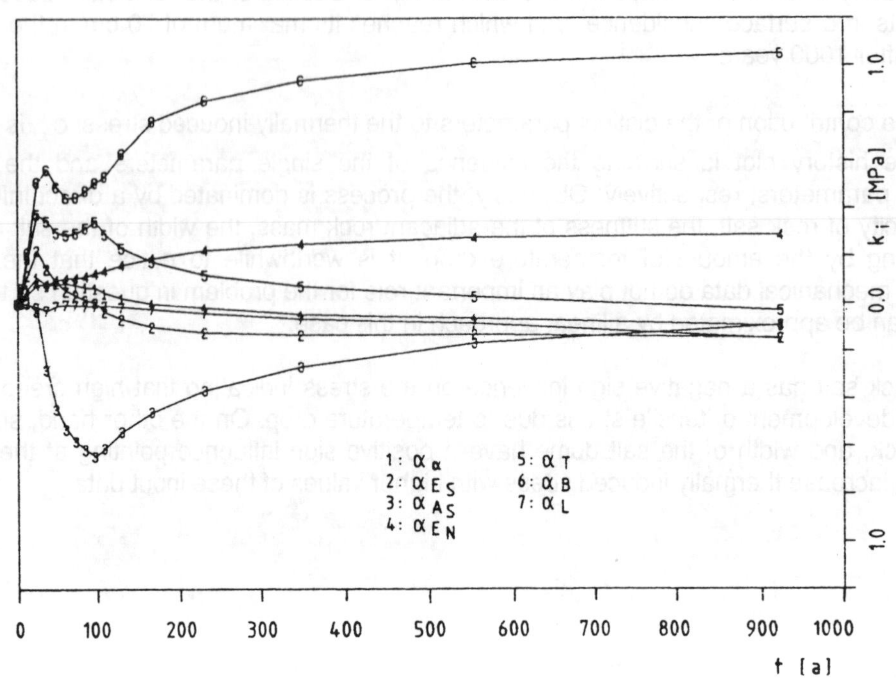

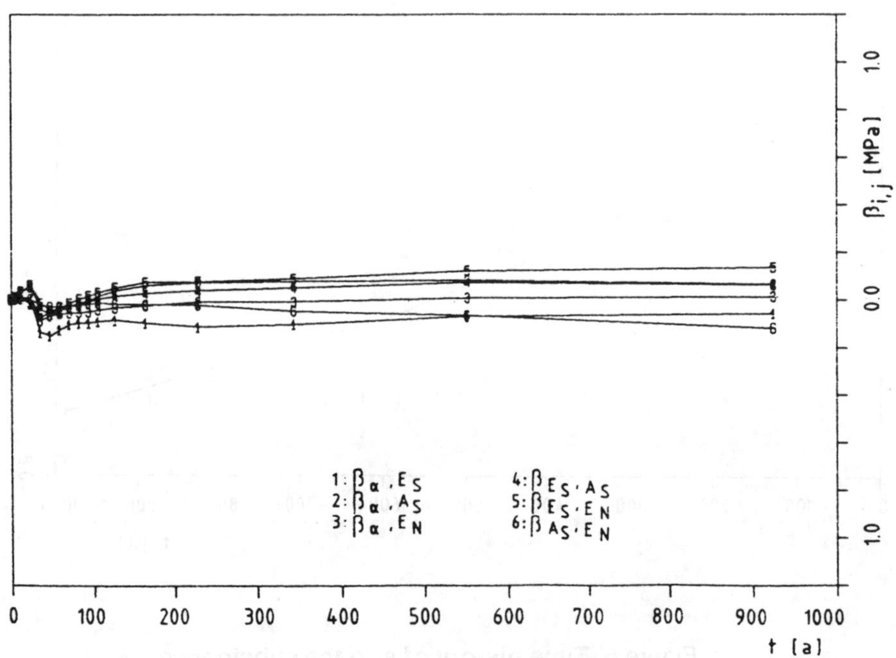

Figure 6. **Time history of sensitivity; all considered parameters with respect to horizontal stress σ_x at the top of the salt dome**

6 Conclusion

Modern concepts of demonstrating geotechnical safety are based on model computations in which all relevant risks are carefully analyzed. Suitable actions, to be taken to prevent those risks, are also investigated by model computations. The assessment of safety, however, requires the consideration of uncertainties in the geotechnical model originating e.g. from the variability of genuine scattered geomechanical data on one side and the variety of design features on the other.

Probabilistic modeling occurs as an efficient tool for evaluating consequences of uncertain future events. The application of probabilistic finite element computations for consequences of a severe temperature drop on the integrity of the salt barrier above a radioactive waste repository shows that probabilistic modeling goes beyond common parametric study because it allows:

- evaluating the significance of uncertainties,
- interpreting the geomechanical behavior on the basis of sensitivities, and
- assessing of complex input parameter interrelation.

As an example, impact of a temperature decrease during glacial age on the integrity of the salt barrier above a repository has been analyzed. A sensitivity study showed that besides the depth of the salt dome also the shape of the dome and the stiffness of the adjacent rock have a significant influence on the development of tensile stress at the top of the salt dome.

However, creep capacity of rock salt enables the salt formation to reduce the total amount of tensile stress significantly. Even under very conservative assumptions, thermally induced fractures will only develop as far as about 100 meters depth, and therefore will not endanger the barrier function.

With respect to the computational procedure, it has to be mentioned that the linear approach to the system response surface is valid for the case in question but can be insufficient for other evaluations. A nonlinear approach would be recommendable. Respective development is under progress.

7 Acknowledgment

This work is based on research which has been supported by the Federal Ministry for Research and Technology, BMFT, under contract 02 E 8431 3 -"Accuracy of thermomechanical 3D-model computations in rock salt".

8 References

[1] IAEA - International Atomic Energy Agency: Safety Principles and Technical Criteria for the Underground Disposal of High Level Radioactive Wastes, Safety Series No.99, Vienna (1989)

[2] Kaplan, S. and Garrick, J.: "On the quantitative definition of risk", Risk Analysis, 1,1 (1981)

[3] CEN - European Committee for Standardization: "Eurocode 7 Part 1,Geotechnical Design, General Rules", working document

[4] Liu, W.K. and Belytschko, T. (editor): "Computational Mechanics of probabilistic and reliability analysis", Elmpress International, Lausanne, Washington (1989) 622

[5] Dong, Z., Eickemeier, R. and Wallner, M.: "Probabilistic Structural Analysis of Thermo-mechanical Response in Rock Salt", presented at 3rd Conf. on The Mechanical Behavior of Salt, Paliseau 1993, preprint
[6] Cochran, W.G., and Cox,G.M.: "Experimental Design", John Wiley & Sons, Inc., New York (1957).
[7] Bauer,G.: "Kryogene Klüfte in norddeutschen Salzdiapiren (Temperature induced fractures in salt domes in Northern Germany)"
[8] Wallner M: "Stability Demonstration Concept and Preliminary Design Calculations for the Gorleben Repository", Proc. Waste Management, Tuscon, Vol II, (1986), 145-151
[9] Delisle, G.: "Permafrost in Nordeuropa und Genese glazialer Rinnen (Perma-frost in North Europe and development of glacial induced channels)", BGR-report (1988), archive-no.: 103486
[10] Morgen, H.S., Wallner, M. u. Munson, D.E.: "Result of an International Parallel Calculations Exercise Comparing Creep Responses Predicted With Three Computer Codes for Two Excavations in Rock Salt", SANDIA REPORT, SAND 87-2125. UC-70, (1987) 72

APPENDIX A: **System Response Approach**

The system response y is considered to be a function of place x, time t and input variable a_j with distinct parameters A_j:

$$y = f(x, t, a_1, \ldots, a_n)$$

with
$$a_j \in \{A_j^{(1)}, A_j^{(2)}, \cdots, A_j^{(L)}\},\ j = 1, 2, \ldots, n.$$

If we assume a linear approach to the system response surface, taking into account interrelations of the input variables, the approximation becomes:

$$y \approx \hat{y} = \alpha_0 + \sum_{i=1}^{n} \alpha_i z_i + \tfrac{1}{2} \sum_{i=1}^{n} \sum_{j=1}^{n} (1 - \delta_{ij}) \beta_{ij} z_i z_j$$

with
$$\delta_{ij} = 0 \text{ for } i \neq j,$$
$$= 1 \text{ for } i = j,$$

$$\beta_{ij} = \beta_{ji},$$

and
$$z_i = \frac{2a_i - A_i^{(1)} - A_i^{(L)}}{A_i^{(L)} - A_i^{(1)}}$$

for normal distributed input variables.

The sensitivity with respect to the j-th parameter is the derivative of the system response and can be approximated by:

$$Sens_j = \left[\alpha_j + \sum_{i=1}^{n} (1 - \delta_{ij}) \beta_{ij} z_i \right] \frac{\partial z_j}{\partial a_j}$$

with
$$\frac{\partial z_j}{\partial a_j} = \frac{2}{A_j^{(2)} - A_j^{(1)}}.$$

The system response y can be approximated by $\hat{y}$:

$$\hat{y}(a_1, \ldots, a_n) = \alpha_0 + \sum_{i=1}^{n} \alpha_i z_i + \tfrac{1}{2} \sum_{i=1}^{n} \sum_{j=1}^{n} (1 - \delta_{ij}) \beta_{ij} z_i z_j$$

If we consider the input parameter random-variables whose probability density functions $\{f_j(a_j)\}_{j=1}^{n}$ are given, the system response Y is also a random-variable. In order to find the cumulative distribution function of the system response Y, one can take $\hat{Y}$ as an approximation to Y:

$$\hat{Y}(A_1,\ldots,A_n) = \alpha_0 + \sum_{i=1}^{n} \alpha_i Z_i + \tfrac{1}{2}\sum_{i=1}^{n}\sum_{j=1}^{n}(1-\delta_{ij})\beta_{ij} Z_i Z_j$$

The probability of $P\{y^{(0)} \le Y \le y^{(1)}\}$ can be theoretically calculated if we assume the system response not to be dependent on interrelations of the input parameters. If all input parameter are normal distributed, i.e. $A_j \approx N(\mu_j, \sigma_j^2)$, j = 1,2,...n, we have then for the *expected value* $E(\hat{Y})$ and the *standard deviation* $\sigma^2(\hat{Y})$ the following expressions:

$$E(\hat{Y}) = \alpha_0 + \sum_{j=1}^{n} \alpha_j E(Z_j)$$

$$\sigma^2(\hat{Y}) = \sum_{j=1}^{n} \alpha_j^2 Var(Z_j)$$

with:

$$E(Z_j) = \frac{2}{A_j^{(L)} - A_j(1)}\left(\mu - \frac{A_j^{(1)} + A_j^{(L)}}{2}\right)$$

and

$$Var(Z_j) = \left(\frac{2}{A_j^{(L)} - A_j^{(1)}}\right)^2 \sigma_j^2.$$

In this case the *probability distribution* of $\hat{y}$ results in:

$$P\{y^{(0)} \le \hat{y} \le y^{(1)}\} = \Phi\left(\frac{y^{(1)} - E(\hat{y})}{\sigma(\hat{y})}\right) - \Phi\left(\frac{y^{(0)} - E(\hat{y})}{\sigma(\hat{y})}\right)$$

and the *cumulative distribution function (CDF)* of $\hat{y}$ is given by:

$$P\{\hat{y} \le y\} = \Phi\left(\frac{y - E(\hat{y})}{\sigma(\hat{y})}\right)$$

which approximates the equivalent *CDF* of *Y*. Φ is the *standard normal cumulative distribution function.*

APPENDIX B: **Orthogonal matrix**

The parametric study is based on a design matrix X. This matrix X is called orthogonal, if $X^T X$ is a diagonal matrix.

The normalized orthogonal matrix for:

- the 4 geomechanical input data, namely stiffness E_S, creep A_S, and thermal expansion α of rock salt, stiffness E_N of the adjacent rock,
- the 2 geometrical input data, namely width *B*, and length *L* of the salt dome,
- temperature decrease *T*, and
- the specified interrelations between the geomechanical data

can easily be determined as compiled in the following table:

	α	E_S	A_S	E_N	*T*	*B*	*L*	α E_S	α A_S	α E_N	E_S A_S	E_S E_N	A_S E_N
1	-	-	-	-	-	-	-	+	+	+	+	+	+
2	-	-	-	+	+	-	+	+	+	-	+	-	-
3	-	-	+	-	+	+	-	+	-	+	-	+	-
4	-	-	+	+	-	+	+	+	-	-	-	-	+
5	-	+	-	-	-	+	+	-	+	+	-	-	+
6	-	+	-	+	+	+	-	-	+	-	-	+	-
7	-	+	+	-	+	-	+	-	-	+	+	-	-
8	-	+	+	+	-	-	-	-	-	-	+	+	+
9	+	-	-	-	+	+	+	-	-	-	+	+	+
10	+	-	-	+	-	+	-	-	-	+	+	-	-
11	+	-	+	-	-	-	+	-	+	-	-	+	-
12	+	-	+	+	+	-	-	-	+	+	-	-	+
13	+	+	-	-	+	-	-	+	-	-	-	-	+
14	+	+	-	+	-	-	+	+	-	+	-	+	-
15	+	+	+	-	-	+	-	+	+	-	+	-	-
16	+	+	+	+	+	+	+	+	+	+	+	+	+

The negative sign is representing z_i = -1, and the positive sign z_i = +1.

Long Term Mineralogical Changes in Salt Formations Due to Water and Brine Interactions

H.-J. Herbert and **W. Brewitz**

GSF-Institut für Tieflagerung, Germany

Abstract

Four very common long term mineralogical changes in salt formations are discussed in the view of the safety considerations for underground repositories. Two of these processes, the "Hartsalz" and "Carnallitit" dissolution were studied in two large scale in situ experiments. The results are presented and compared with the results of the geochemical modelling with the computer code EQ3/6. Furthermore the reactions leading to the formation of the gypsum cap rock on the top of the Zechstein salt formations and to the polyhalitization of anhydrite are discussed. Geological field observations and mineral assemblages agree well with the results of the geochemical modelling employing the Pitzer formalism along with the Harvie, Møller and Weare database.

We conclude that once the mechanisms of the chemical reactions are well understood it becomes possible to evaluate realistically whether such processes, when encountered in the repository, are still active or whether they are finished. It also becomes possible to estimate the volume changes associated with the reactions and thus the impact of these reactions on the integrity and the geomechanical stability of the salt formation. The intimate knowledge of the reaction mechanisms of the short and long term changes in the mineralogical assemblages and the associated brine chemistry is a first prerequisite for the correct evaluation of the origin of brines. Thus, it is essential for the correct assessment of the hazards which brine inflows may pose for the safety of a repository in salt formations.

1. Background

In the safety assessment of a repository in salt formations - according to the German concept -the investigation of the hypothetical accident scenario of water or brine inflow is of particular significance. The water path is considered to be the most important way for the mobilisation and release of radionuclides from the repository. The volume and composition of the contaminated brines is largely dependant on the accident scenario which is considered in the safety assessment.

Two different accident scenarios are discussed in Germany:

- the limited scenario with a volume of maximum 1000 m^3 of intruding brine
- the unlimited scenario with a brine volume equal to the volume of the repository (complete flooding)

The results of the safety assessment with these two different scenarios differ significantly. Therefore good reasons are needed for the selection of the most realistic scenario. The brine volume which has to be considered depends mainly on the origin of the brines and the access ways into the repository. Brines genetically linked to the origin of the salt formation generally have a small volume, whereas brines from the overburden have an unlimited reservoir and can completely flood the repository, provided that access ways exist. The knowledge of the origin of brines provide informations on possible pathways within the salt formation and possible connections to the overburden.

Brines formed in the geological processes which led to the formation and partial transformation of the salt formation have not only a relatively small volume but generally are trapped within the salt formation. Such brines have no connection with the overburden. They can enter the repository if they are encountered during the exploration and construction phase of the repository. But as they have no connection with the overburden their reservoir is restricted. No major brine inflow has to be expected. The practical experience made in the salt and potash mining industry shows that such brines generally dry up after a short period of inflow. If they are encountered during the operational phase of the repository they can easily be collected and removed. Once they have dried up no further brines from that reservoir must be expected. Such brines are harmless.

The situation is completely different if the intruding brines come from the overburden. Generally they don't dry up. Often they dry up in one place and reappear quickly in another place in the mine. Eventually they may completely flood the repository. Such brines are dangerous. It is essential to have tools at hand to distinguish such brines from harmless brines. Often the two different types of brines have the same chemical composition.

This discussion shows how essential for the safety of a repository it is to collect reliable information on the origin of brines encountered during the exploration phase of the repository. In addition the discussion illustrates the limited value of the limited accident scenario in the safety assessment. It is simply not possible to demonstrate that using the limited scenario is a conservative approach.

Geomechanical processes may lead to an inflow of harmless brines with a small reservoir in the postoperational time of the repository. In this case the limited scenario would be the more realistic one. But it can not be excluded that geomechanical processes can cause the inflow of large volumes of brines from the overburden. This would require the consideration of the unlimited accident scenario. As no supervision of the repository in the postoperational time is required by the German concept such processes and the intrusion of brines would occur unnoticed. As nobody can tell in this case which kinds of brines will enter the repository the only conservative approach is the unlimited accident scenario.

What has this scenario discussion to do with the topic of this conference - the "long term geological changes"? Can we learn enough from the investigation of the long term changes about the evolution and consequently about the origin of brines to be able to make useful contributions to the scenario discussion? The answer is - yes, we can ! What we need is a thorough understanding of the geochemical reactions, in order to be able to distinguish during the operational time of a repository between hazardous brines from the overburden and harmless brines from the interior of the salt formation.

In the following chapters the results of two large scale in situ experiments of relevant water rock interactions in salt formations, "Hartsalz" and Carnallitit" dissolution (chapters 2.1 and 2.2) are discussed and compared with results of the geochemical modelling with EQ3/6, using the Pitzer-formalism and the Harvie Moller, Weare (1984) database. Two further long term geological processes, the "Gypsum cap rock formation" (chapter 2.3) and the "Polyhalitisation of anhydrite" (chapter 2.4) are investigated theoretically via geochemical modelling of the involved reactions.

2. Short and long term water rock interactions in salt formations

2.1. "Hartsalz" dissolution (H_2O + sylvite + kieserite + halite + anhydrite)

"Hartsalz" is the german term for a quite common and valuable secondary potash formation within the Zechstein salt sequences in Northern Germany. There are different types of "Hartsalz", like kieseritic, anhydritic and langbeinitic "Hartsalz". The main components of all "Hartsalz" types are sylvite and halite. Minor components are different sulfates like kieserite, anhydrite and langbeinite which give the name of a specific "Hartsalz" type. All "Hartsalz" types were formed by the interaction of water with the primary potash formation "Carnallitite" which in turn was formed by the primary precipitation from the evaporating sea water. As the kieseritic "Hartsalz" is the most common "Hartsalz" type in the Zechstein salt formations the dissolution behaviour of this type was studied in detail.

A large scale in-situ experiment was carried out in the potash mine Hope, near Hannover in Northern Germany. In the Hope mine "Hartsalz", consisting of 61.4% halite, 31.7% sylvite, 3.4% kieserite and 1.7% anhydrite was mined. The abandoned mine was flooded with a NaCl saturated brine. A research programme was set up to record and evaluate data on the geochemical, geomechanical and geophysical processes occuring during and after flooding. The experimental results of the geochemical programm which monitored the changes in the chemical composition of the charged NaCl solution in contact with "Hartsalz", were published by Herbert and Reichelt (1992). A more complete evaluation of the results and a detailed comparison between experimental data and geochemical modelling was carried out later by Herbert. Some of the results of this work are illustrated in figures 1 - 3. Figure 1 shows the chemical evolution of the brine, the amount of rock affected by the dissolution process and the time needed for the solution to reach different concentrations. The symbols represent samples taken at different times from different locations in the mine, whereas the lines show the evolution as calculated with the geochemical code EQ3/6. The close resemblance between calculated and experimental results is noticeable. This demonstrates the great internal consistency of the employed thermodynamical database from Harvie, Moller and Weare (1984) with respect to the Pitzer coefficients and the mineral solubility constants. Furthermore the good agreement of the experimental and theoretical data demonstrates that all minerals which play a substantial role in the studied reaction have been taken into account in the calculation , thus demonstrating that the "Hartsalz" dissolution is well understood.

The vertical line in figure 1 marked with R1 indicates the reaction step for which a quantitative reaction comprising the initial solution, the resulting solution, all reactants and all products, was calculated using the results of the geochemical modelling. As we know the composition of the initial and resulting solutions as well as the reactants involved in the reaction and considering the good agreement between experimental data and computer modelling at each step of the reaction, we conclude that we can extrapolate the results reliably beyond the limits of our experimental data. Under this premise we can show that the experimental results cover only a small part of the entire reaction. The line R1 in figure 1 corresponds to R1 in figure 2. The lines R1 - R7 indicate the steps for which quantitative reactions were calculated. Figure 3 demonstrates, that during the reaction water is continuously consumed and the reaction terminates when no water is left. The water is consumed during the reaction primarily via the formation of the mineral kainite. During the reaction (R7, figure2) 1 kg of water in the initial solution can theoretically affect about 90 kg of "Hartsalz". About 27 moles of sylvite, 22 moles of kieserite and 11 moles of anhydrite can be dissolved and 5.5 moles of carnallite, 5.7 moles of halite, 11 moles of kainite and 5.5 moles of polyhalite can precipitate. Between R6 and R7 the chemical composition of the solution does not change anymore. It has reached the invariant composition of an IP 21 solution in the six component system Na-K-Ca-Mg-Cl-SO_4. This solution is saturated with halite, sylvite, carnallite, kainite and polyhalite. In spite of the solution staying constant in composition between R6 and R7, the dissolution continues, as the solution is not yet saturated with kieserite and anhydrite. The dissolved amounts of $CaSO_4$ and $MgSO_4$ are consumed by the formation of new kainite. This process finaly stops when no water is left.

Whereas in flooded "Hartsalz" mines the resulting solutions normally reach the composition of an IP21 solution (at R6 in figure 2), it is very unlikely that the reaction will proceed under natural conditions to the very end (R7 in figure 2). The reaction will stop whenever one of the reactants is exhausted or has no more direct contact with the solution. Therefore it is very important to have natural analogues, i.e. flooded potash mines, in order to be able to show which steps of the theoretical reaction can be reached under natural conditions (different mineralogical compositions of affected potash rocks, different outcrops etc.). It is equally important to have a thorough theoretical understanding of the expected reactions.

2.2. "Carnallitit" dissolution (H_2O + carnallite + kieserite + halite + anhydrite)

"Carnallitite", a potash rock consisting mainly of carnallite, kieserite and halite, is very common in the Zechstein salt sequences in Northern Germany. Its dissolution behaviour, which is very different from that of "Hartsalz", has been studied in detail as well. Whereas "Hartsalz" is a secondary potash formation, "Carnallitite" was formed by the simultaneous precipitation of carnallite, kieserite and halite from the evaporating sea water. The rock "Carnallitite" can contain minor amounts of sylvite, anhydrite and clay minerals.

The dissolution of "Carnallitite" was studied in a vertical borehole drilled into the "Carnallitit" formation in the Asse Mine, near Braunschweig. The borehole, 40 cm in diameter and 4 m deep, was filled with a NaCl saturated brine. The change of the chemical composition of the brine was meassured for about 4.5 years. The experimental results as well as the results of the geochemical modelling are shown in figures 4 and 5. As in the "Hartsalz" experiment, a good agreement between experimental and theoretical data was obtained. In the "Carnallitit" dissolution experiment the boundary conditions were more favorable, which led to a more complete coverage of the theoretical curves with experimental data. i.e. most of the total reaction path was covered by the experiment. As we have a good agreement between the two sets of data, this experiment is even more valuable in terms of demonstrating the ability of the EQ3/6 code to reliably predict the dissolution reactions in salt formations.

For the safe disposal of hazardous wastes in salt formations it makes quite a difference if "Hartsalz" or "Carnallitit" is present in the openings of the repository. Whereas in the dissolution reaction of "Hartsalz" the initial amount of water is gradually consumed by the formation of hydrated minerals in the dissolution reaction of "Carnallite" the initial amount of water increases (compare figures 3 and 5). The line R4 in figure 4 marks the composition of an invariant solution of the six component system Na-K-Ca-Mg-Cl-SO_4 called IP19. This solution is saturated with halite, carnallite, kieserite, kainite and polyhalite. It is not yet saturated with anhydrite. The saturation with anhydrite is reached at line R5.

Beside the differences in the dissolution behaviour of the two potash formations "Hartsalz" and "Carnallitit" there are also similarities. The resulting solutions IP 21 and IP 19 are saturated with halite, carnallite, kainite and polyhalite and have consequently very high Mg concentrations (above 100.000 mg Mg/kg H_2O).

2.3. Gypsum cap rock formation

A gypsum cap rock exists at the top of every salt structure in Northern Germany. This rock is formed by a hydration process of anhydrite (gypsification) due to the dissolution of the salt formation by groundwater. During this process halite is dissolved and removed as NaCl brine with the ground water flow, whereas the anhydrite with a much lower solubility is hydrated and transformed into gypsum. From the thickness of the gypsum cap rock the dissolution rate of the salt structure can be approximated. This is an important number for the location of the repository at a safe depth within the salt formation. The thickness of the gypsum cap rock at the top of the Asse salt structure varies between 12 and 17 m (Batsche and Klarr 1979). From the Gorleben site Bornemann and Fischbeck (1986) reported cap rock thicknesses between 0 and 111 m with a very accentuated relief. They calculated an average dissolution rate of the salt structure of 0.04 mm/a. In an isotope study, Herbert et al. (1992) determined the average temperature during the Gorleben cap rock formation to be 6°C. In this study the geochemical processes during the cap rock formation were studied, in order to get informations about the quantities of water, of reactants (halite and anhydrite) and products (gypsum) involved in the reaction.

The solubility of gypsum and anhydrite in water and NaCl solutions has been investigated by many authors. Literature reviews and new solubility data are reported by Blount and Dickson (1973) and Raju and Atkinson (1990). The reaction H_2O + halite + anhydrite presented in this paper was calculated for the temperature of 6°C with EQ3/6 using the Pitzer database. The results are presented in figures 6 - 10 and tables 1 - 4. The EQ6 reaction path was started with an NaCl saturated solution. The evolution of the brine chemistry is given in table 1. Table 2 shows the consumed reactants and the precipitated products during the reaction. In table 3 the evolution of the mineral saturation states of reactants and products are listed. Table 4 gives the complete reactions at two significant stages of the reaction, which are marked in the figures with lines and indexes R1 and R2.

From the results of the EQ3/6 calculation we conclude, that during the reaction anhydrite is dissolved continuously over a long time range and equal amounts of gypsum and small amounts of halite are precipitated. Therefore only minor changes in the chemical composition of the solution occur. This is illustrated also by the almost constant activity of water in this range of the reaction. When about 90% of the initial water is used up by the formation of gypsum the velocity of anhydrite dissolution exeeds the velocity of gypsum formation. During this phase of the reaction the anhydrite saturation (affinity) in the brine rises. The reaction stops when complete anhydrite saturation is reached. At this moment all the water is consumed by the gypsum formation.

We conclude that the cap rock formation is well understood. Geological, mineralogical and geochemical observations corroborate each other. The formation of the gypsum cap rock can not endanger the safety of a repository, as long as the repository is located deep enough within the salt structure.

2.4. Polyhalitization of anhydrite

The partial or total polyhalitisation of anhydrite can be observed in many salt structures in Northern Germany. The polyhalitisation can be explained by the reaction of a $MgCl_2$-rich, carnallite saturated solution with anhydrite. For the safety of a repository in salt formations it is important to find out by field studies (exploration) whether or not the Hauptanhydrit (A2), a several m thick anhydrite bed in the Zechstein 2, is affected by polyhalitisation. As the molar volume of polyhalite is much higher than that of anhydrite the crystallisation of polyhalite in a closed system may lead to high crystallisation pressures, which can fracture the anhydrite and release brines into the repository. Therefore if polyhalitisation can be observed, the presence or absence of brines in the A2 needs to be determined. The practical experience of the salt mining industry shows that the Hauptanhydrit (A2) quite often is a reservoir for $MgCl_2$ rich, carnallite saturated solutions. The reaction mechanism of this possible long term geological process must be determined qualitatively and quantitatively.

The results of the calculated reaction at 25 °C (figures 11 - 14 and tables 5 - 8) compare well with observed mineral assemblages in different potash mines. However the amounts of anhydrite which can be dissolved in this reaction are extremely small. If the polyhalitisation of large parts of the Hauptanhydrite (A2) is due to the described reaction at low temperatures (around 25°C), extremely high volumes of solutions are involved. However, at such low temperatures there are no other reactions which result in pure polyhalite. Therefore it is likely that the described reaction was involved. Nevertheless the possibility of other reactions leading also to pure polyhalite (and possibly $CaCl_2$-rich solutions) must be investigated.as well.

3. Conclusions

By describing four relevant water rock interactions in salt formations and comparing experimental results with equilibrium calculations (which simulate the long term geological processes) it was demonstrated, that we have a good understanding of these processes. The close correspondence between the calculated and experimental results of the large scale field experiments, the "Hartsalz" and "Carnallit" dissolution, indicates that the geochemical computer code EQ3/6 with the Pitzer-formalism and the Harvie, Moller, Weare (1984) database yield reliable predictions. On this basis the geochemical modelling with EQ3/6 was successfully applied in the investigation of two other long term geological processes in salt formations, the "Gypsum cap rock formation" and the "Polyhalitization of anhydrite". Geological field observations, microscopical observations of mineral assemblages and structures agree well with the results of the geochemical modelling.

We have also drawn the attention to the importance of adequate tools for the characterisation of brines. It must be possible to distinguish brines formed during the long term geological processes (harmless briness trapped in the salt formation) from brines which have achieved their chemical composition from recent water rock interactions. This differentiation is an extremely difficult task as the salt minerals are well soluble and the resulting chemical compositions are the same, no matter how old the water is, or where it comes from. In order to be able to make this distinction on a sound scientific basis it is necessary to understand in detail the nature and the mechanisms of the short term as well as of the long term mineralogical and geochemical changes. EQ3/6 has proved to be a powerful tool in this regard.

It can be assumed that the long term mineralogical changes establish an equilibrium between the brines and the surrounding minerals. Such equilibria can be calculated accurately for the six component system Na-K-Ca-Mg-Cl-SO_4 at 25 °C. In spite of the general good agreement of the results of in situ experiments and geochemical modelling, differences are noted in the sulfate saturation of these natural, young brines and the calculated compositions of these brines. In older brines the sulfate supersaturation tends to disappear. Sulfate supersaturation should not be present in very old brines, which are remnants from the long term geological processes. It is worthwhile to investigate if these differences can be used in order to distinguish reliably old from recent brine compositions. It is especially important, therefore to study not only the mineral assemblages of the long term geological processes but also the fluid phases.

From this investigations we conclude that once the mechanisms of the chemical reactions are well understood it becomes possible to evaluate realistically whether such processes, when encountered in the repository, are still active or whether they are finished. It also becomes possible to estimate the volume changes associated with the reactions and thus the impact of these reactions on the integrity and the geomechanical stability of the salt formation. The intimate knowledge of the reaction mechanisms of the short and long term changes in the mineralogical assemblages and the associated brine chemistry is a first prerequisite for the correct evaluation of the origin of brines and essential for the correct evaluation of the hazards which brine inflows may pose for the safety of a repository in salt formations.

4 Literature

Batsche, H. and Klarr, K.(1979): Beobachtungen zur Gipshutgenese. - In Fifth Symposium on Salt, 1, 9-19, published by the Northern Ohio Geological Society, Cleveland, Ohio.

Blount, C. and Dickson, F. (1973): Gypsum-Anhydrite Equilibria in Systems $CaSO_4$-H_2O and $CaSO_4$-NaCl-H2O. - American Mineralogist, 58, p 323-331.

Bornemann, O. and Fischbeck R. (1986): Ablaugung und Hutgesteinsbildung am Salzstock Gorleben. - Z. dt. geol. Ges., 137, p 71-83.

Harvie, C., Moller, N. and Weare, J. (1984): The prediction of mineral solubilities in natural waters: The Na-K-Mg-Ca-H-Cl-SO_4-OH-HCO_3-CO_3-CO_2-H_2O system to high ionic strengths at 25°C. - Geochim. Cosmochim. Acta, 48, 723-751.

Herbert, H.-J., Bornemann, O. and Fischbeck, R (1990): Die Isotopenzusammensetzung des Gipskristallwassers im Hutgestein des Salzstocks Gorleben - ein Nachweis für die elsterzeitliche Bildung der Hutgesteinsbrekzie. - Kali und Steinsalz, 10, 215-226.

Herbert, H.-J. and Reichelt, Chr. (1992): Sieben Jahre Laugenentwicklung im gefluteten Kalibergwerk Hope - Geochemische Messungen und geochemische Modellierung. - Kali und Steinsalz, 11, 44-48.

Pitzer, K. S. (1973): Thermodynamics of Electrolytes I: Theoretical basis and general equations. -J. Phys. Chem. 77, 268-277.

Raju K. and Atkinson, G. (1990): The thermodynamics of "Scale" Mineral Solubilities. 3. Calcium Sulfate in Aqueous NaCl. - Journal of Chemical & Engineering Data, 35, 361-367.

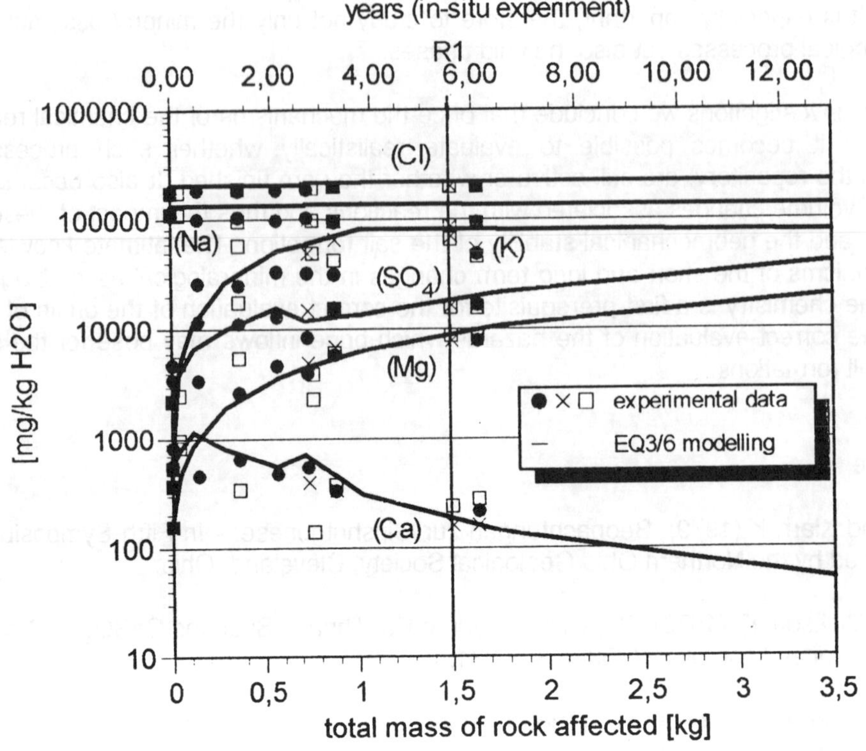

Fig.1 Initial stage of "Hartsalz" dissolution. Comparison between experimental data and the EQ3/6 modelling, 25°C, HMW database.

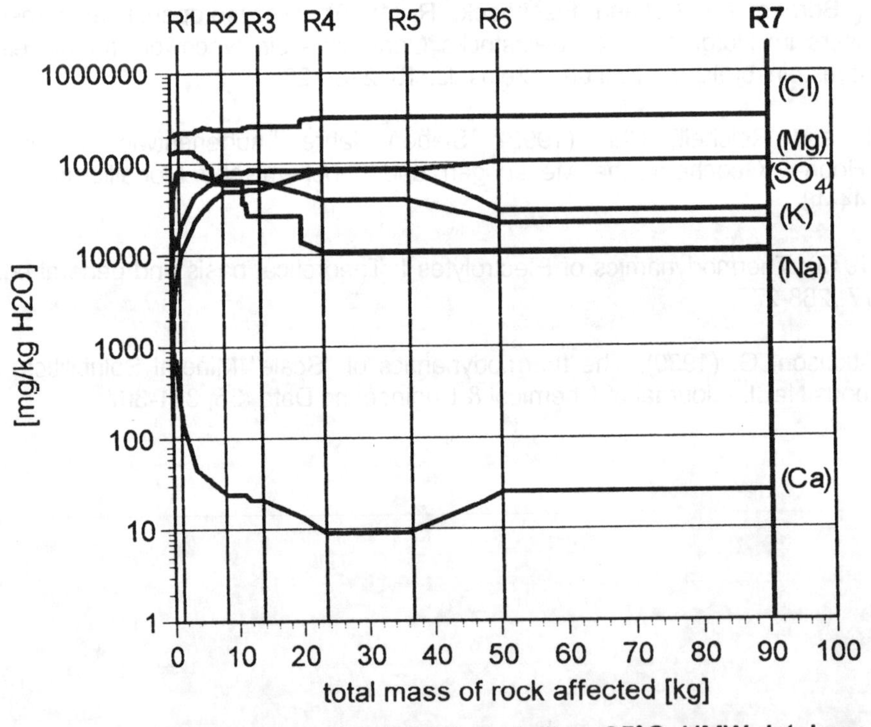

Fig. 2 Dissolution of "Hartsalz". EQ3/6 modelling, 25°C, HMW database.

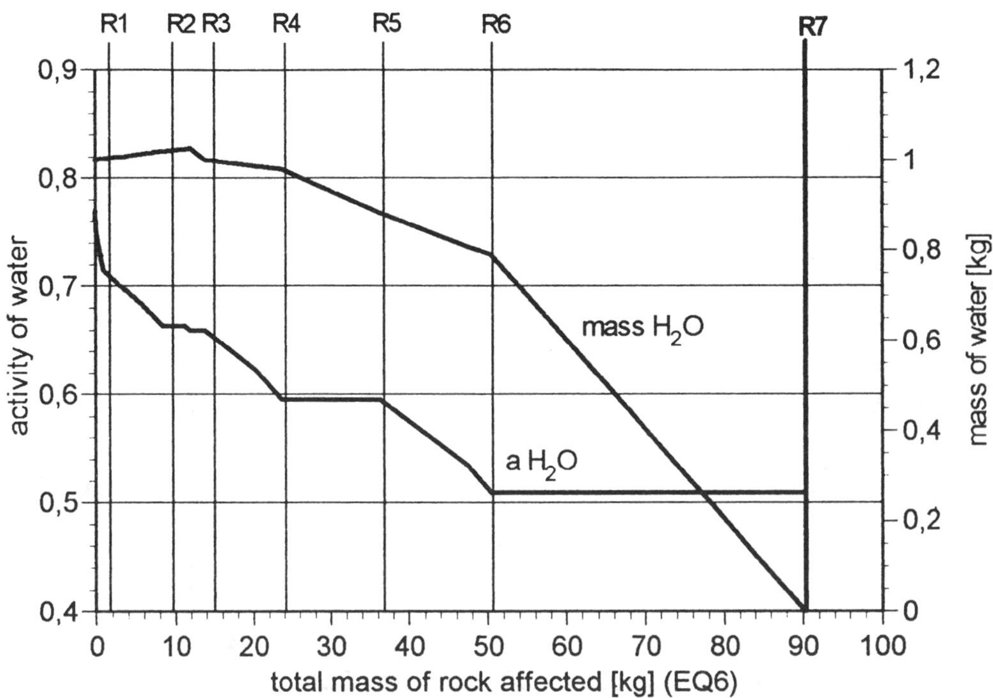

Fig. 3 **Evolution of the total mass of water and the activity of water during the dissolution of "Hartsalz". EQ3/6 modelling, 25°C, HMW database.**

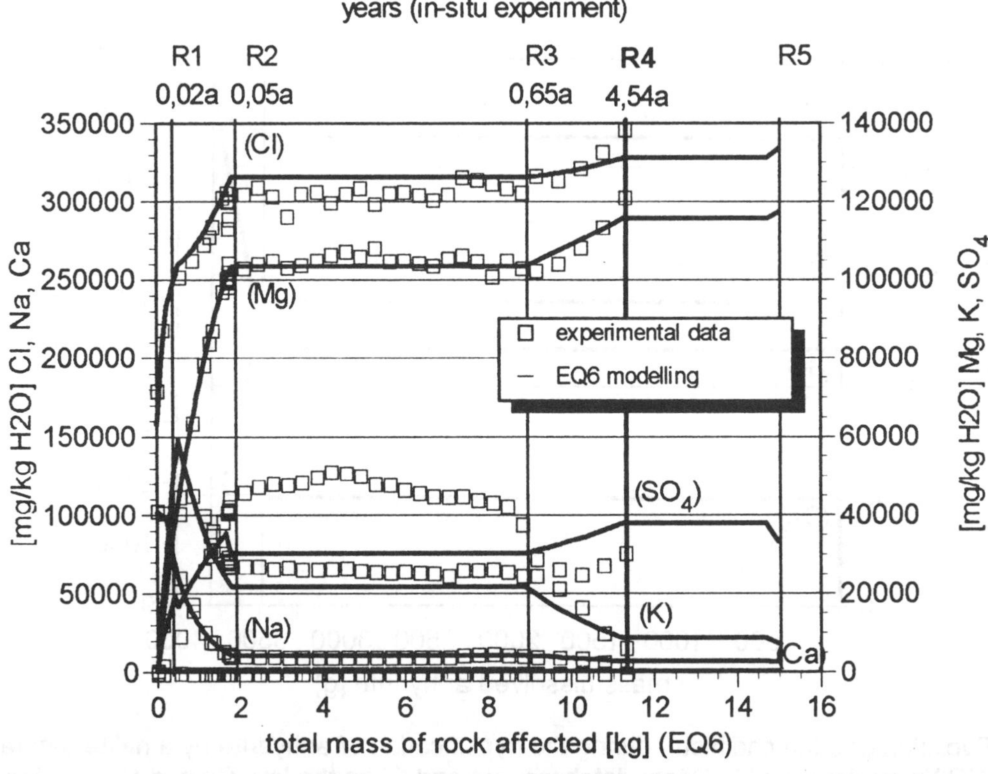

Fig. 4 **Dissolution of "Carnallitite". Comparison between experimental data and the EQ3/6 modelling, 25°C, HMW database.**

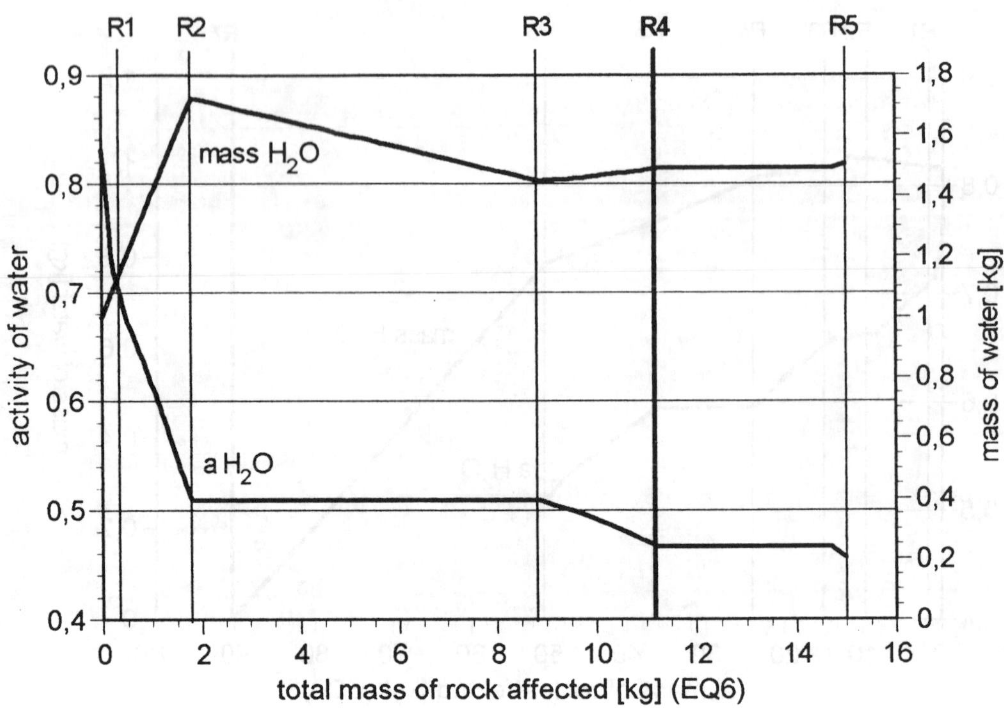

Fig. 5 **Evolution of the total mass of water and the activity of water during the dissolution of "Carnallitite". EQ3/6 modelling, 25°C, HMW database.**

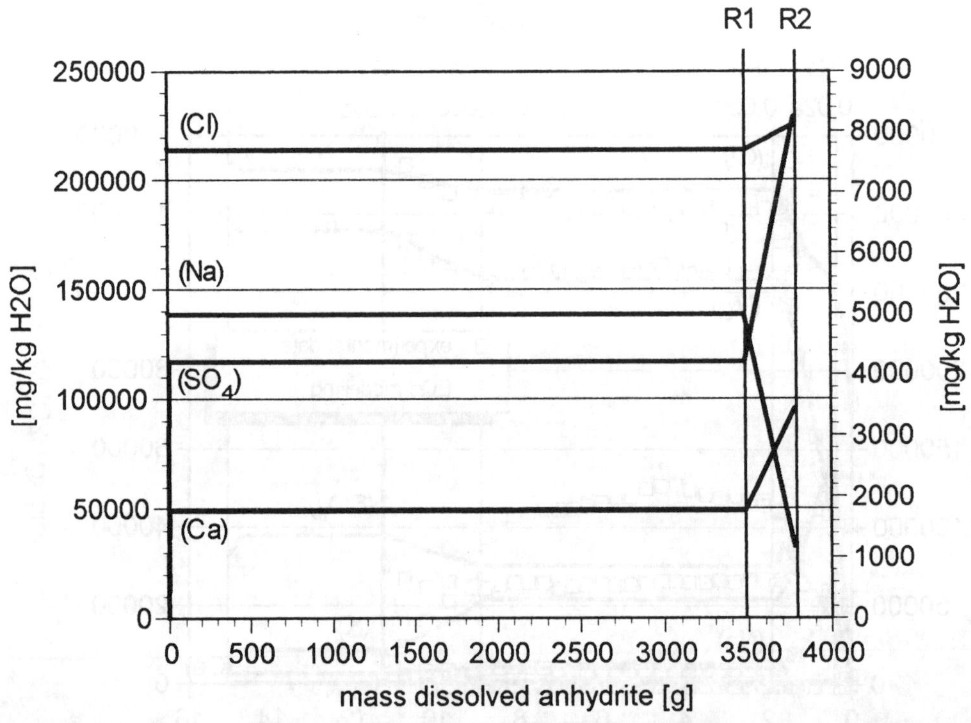

Fig. 6 Evolution of brine chemistry during the gypsification of anhydrite by a halite-saturated solution. EQ3/6 modelling, 6°C, Pitzer database. Na and Cl on the left, Ca and SO_4 on the right y-axis

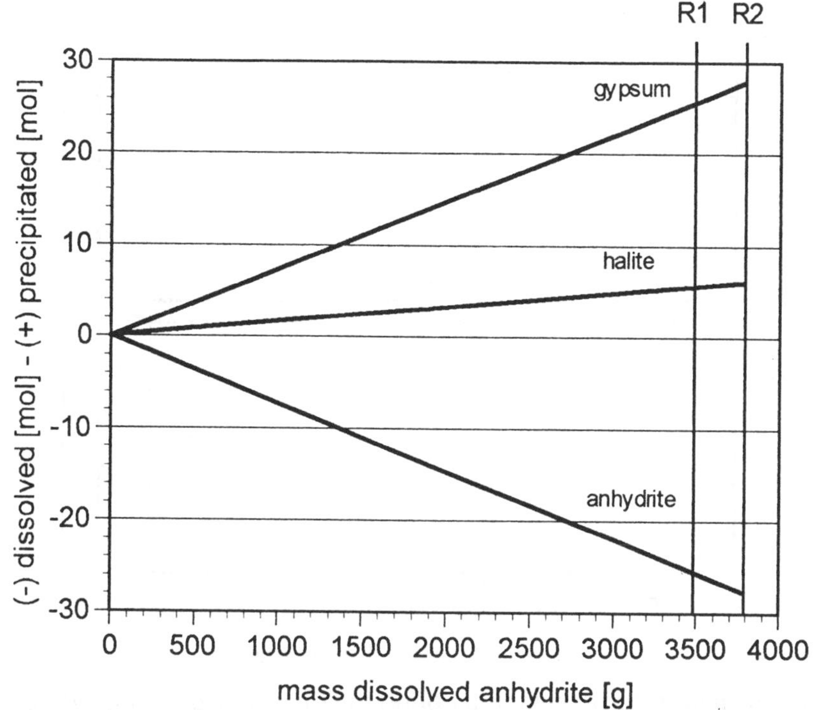

Fig. 7 **Dissolved and precipitated minerals during the gypsification of anhydrite by a halite-saturated solution. EQ3/6 modelling, 6°C, Pitzer database.**

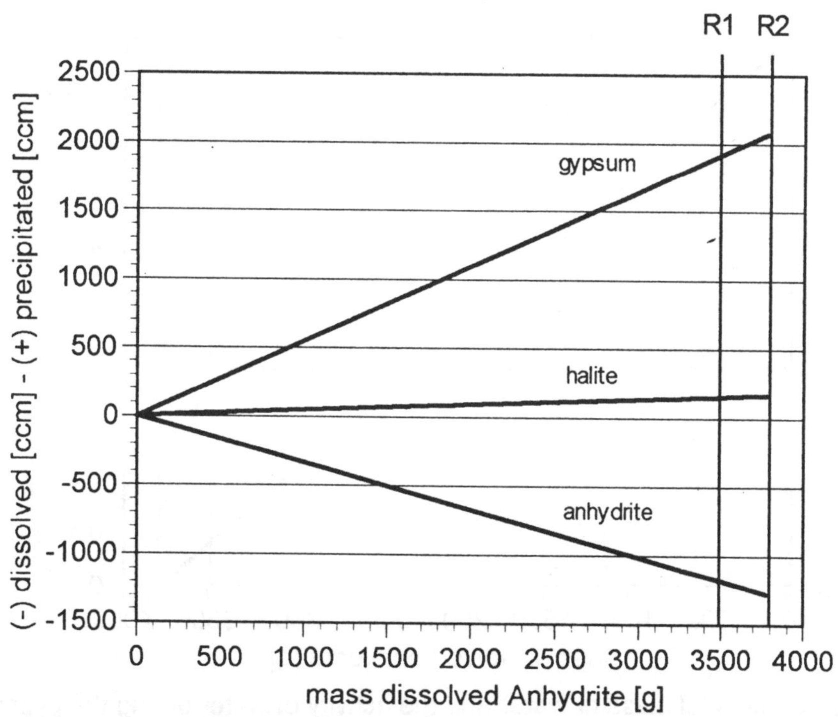

Fig. 8 **Volume changes during the gypsification of anhydrite by a halite-saturated solution. EQ3/6 modelling, 6°C, Pitzer database.**

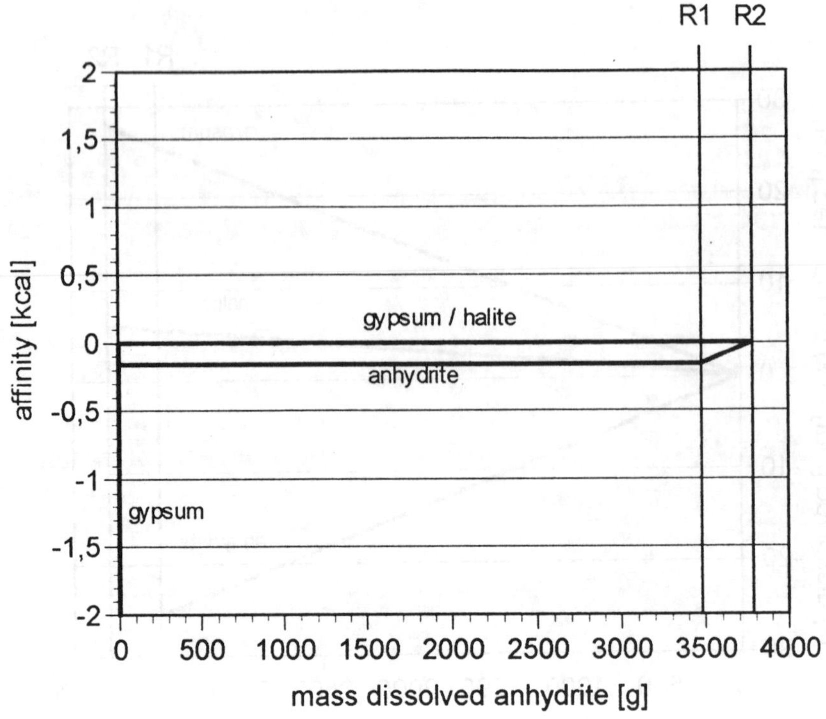

Fig. 9 **Evolution of mineral saturation states during the gypsification of anhydrite by a halite-saturated solution. EQ3/6 modelling, 6°C, Pitzer database.**

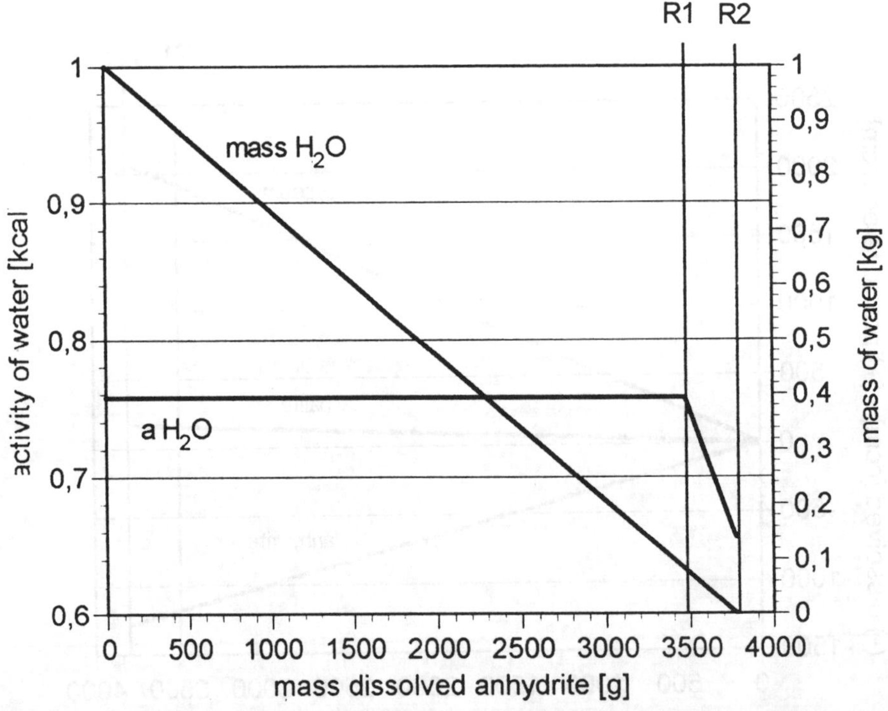

Fig. 10 **Evolution of the total mass of water and the activity of water during the gypsification of anhydrite by a halite-saturated solution. EQ3/6 modelling, 6°C, Pitzer database.**

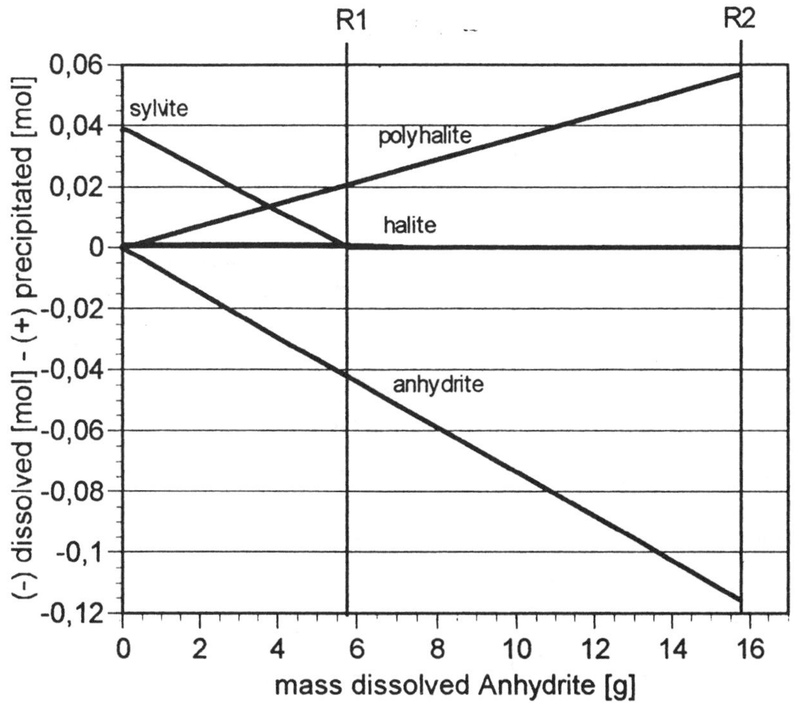

Fig. 11 **Dissolved and precipitated minerals during the polyhalitization of anhydrite by a carnallite-saturated solution (Q). EQ3/6 modelling, 25°C, HMW database.**

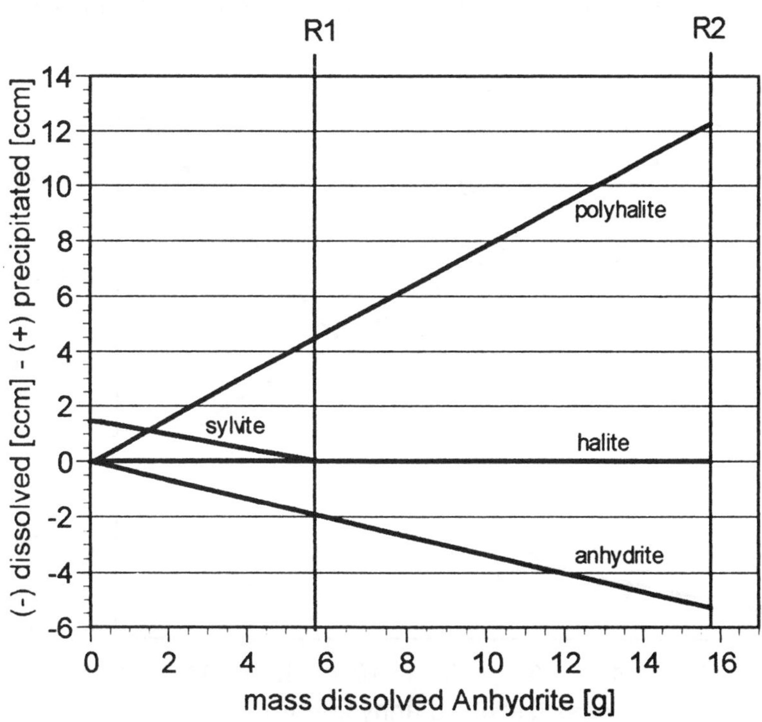

Fig. 12 **Volume changes during the polyhalitization of anhydrite by a carnallite-saturated solution (Q). EQ3/6 modelling, 25°C, HMW database.**

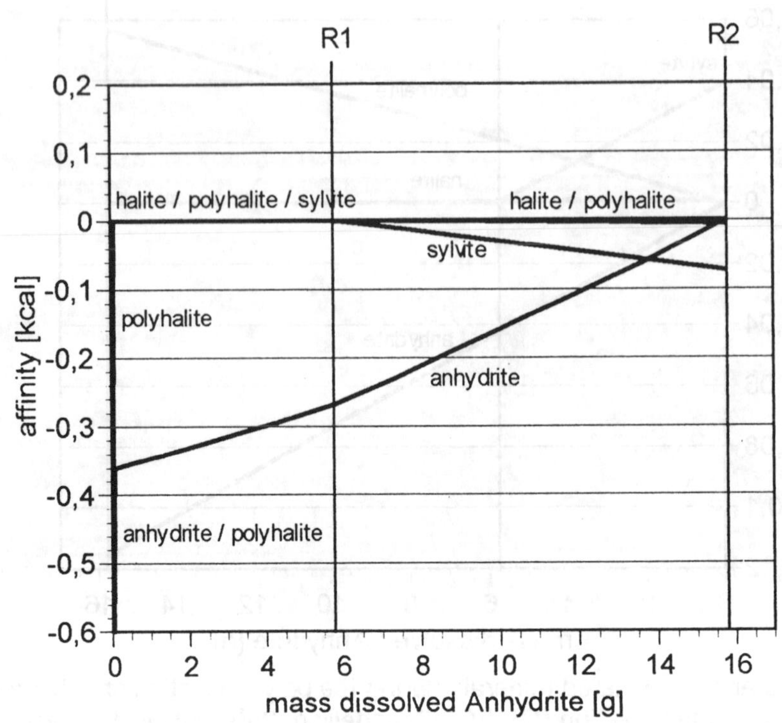

Fig. 13 **Evolution of mineral saturation states during the polyhalitisation of anhydrite by a carnallite-saturated solution (Q). EQ3/6 modelling, 25°C, HMW database.**

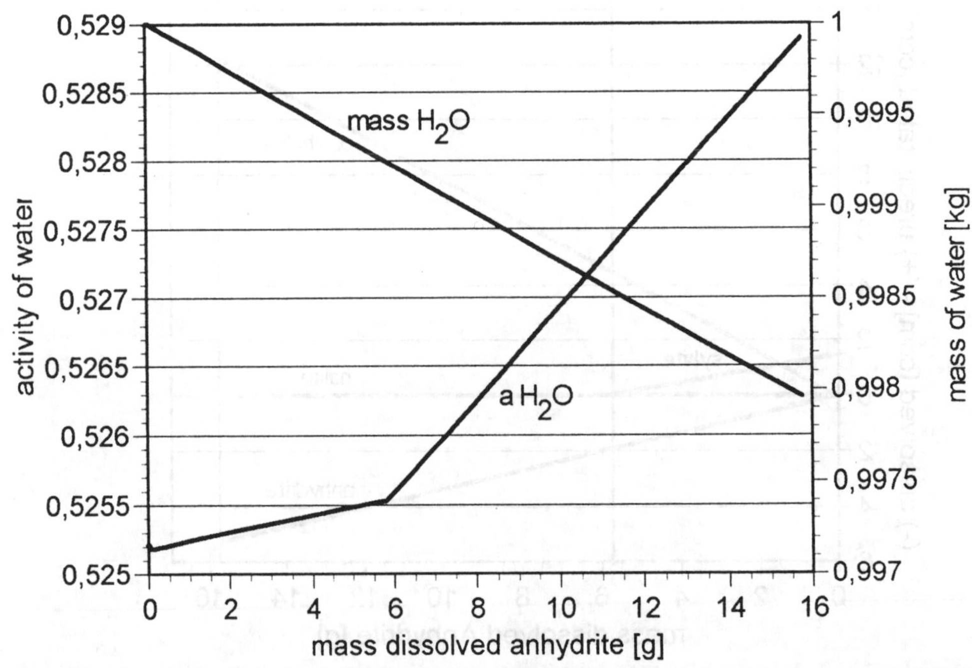

Fig. 14 **Evolution of the total mass of water and the activity of water during the polyhalitisation of anhydrite by a carnallite-saturated solution (Q). EQ3/6 modelling, 25°C, HMW database.**

Table 1. **Evolution of brine chemistry during the gypsification of anhydrite – cap rock formation – by a NaCl-saturated solution. EQ3/6 modelling, 6°C, Pitzer database.**

Element Totals for the Aqueous Solution								
reaction	elements [mol]							
progress	S	O	Ca	Cl	H	K	Mg	Na
6,0628	1,041E-14	55,508	2,4951E-14	6,0605	111,02	2,558E-14	4,114E-14	6,0605
6,0628	9,6211E-06	55,508	9,6211E-06	6,0605	111,02	2,558E-14	4,114E-14	6,0605
6,1066	0,043756	55,683	0,043756	6,0292	111,02	2,558E-14	4,114E-14	6,0292
10	0,037608	47,86	0,037608	5,1821	95,419	2,558E-14	4,114E-14	5,1821
10	0,037608	47,86	0,037608	5,1821	95,419	2,558E-14	4,114E-14	5,1821
20	0,021818	27,765	0,021818	3,0063	55,355	2,558E-14	4,114E-14	3,0063
30	0,0060273	7,6702	0,0060273	0,83051	15,292	2,558E-14	4,114E-14	0,83051
31,623	0,0034648	4,4093	0,0034648	0,47742	8,7908	2,558E-14	4,114E-14	0,47742
33,817	1,7338E-08	1,1282E-05	1,7338E-08	1,2872E-06	2,3425E-05	2,558E-14	4,114E-14	2,881E-07
33,817	1,7336E-08	1,1281E-05	1,7336E-08	1,2871E-06	2,3422E-05	2,558E-14	4,114E-14	2,88E-07

Table 2. **Evolution of reactants and products during the gypsification of anhydrite – cap rock formation – by a NaCl-saturated solution. EQ3/6 modelling, 6°C, Pitzer database.**

Reactant and Product Summary				
reaction	reactants [mol]	products [mol]		
progress	anhydrite	halite	gypsum	anhydrite
6,0628	0	0,002288	0	0
6,0628	9,6211E-06	0,0022948	0	0
6,1066	0,043786	0,033552	3,0152E-05	0
10	3,9372	0,88068	3,8996	0
10	3,9372	0,88068	3,8996	0
20	13,937	3,0565	13,915	0
30	23,937	5,2323	23,931	0
31,623	25,56	5,5854	25,557	0
33,817	27,754	6,0628	27,754	0
33,817	27,754	6,0628	27,754	1,1182E-08

Table 3. **Evolution of the mineral saturation states during the gypsification of anhydrite – cap rock formation – by a NaCl-saturated solution. EQ3/6 modelling, 6°C, Pitzer database.**

Summary of Pure Mineral Saturation States					
reaction	affinities [kcal]				
progress	anhydrite	gypsum	halite	glauberite	$CaSO_4 \cdot 0{,}5\ H_2O$ (beta)
6,0628	-31,876	-31,715	0	-48,351	-33,031
6,0628	-9,456	-9,295	0	-14,479	-10,612
6,1066	-0,16	0	0	-0,538	-1,315
10	-0,16	0	0	-0,538	-1,315
10	-0,16	0	0	-0,538	-1,315
20	-0,16	0	0	-0,538	-1,315
30	-0,16	0	0	-0,538	-1,315
31,623	-0,16	0	0	-0,538	-1.315
33,817	0	0	0	-1,896	-1,196
33,817	0	0	0	-1,896	-1,196

Table 4. **Quantitative reaction equations during the gypsification of anhydrite – cap rock formation – by a NaCl-saturated solution. EQ3/6 modelling, 6°C, Pitzer database.**

Reaction 1	reaction progress 31,623
initial solution	[55,508 O + 111,02 H + 6,0605 Na + 6,0605 Cl]
initial reactants	25,56 $CaSO_4$ (anhydrite)
final products	25,557 $CaSO_4 \cdot 2\ H_2O$ (gypsum) + 5,58311 NaCl (halite)
final solution	[4,4093 O + 8,7908 H + 0,47742 Na + 0,00346 Ca + 0,47742 Cl + 0,00346 S]
Reaction 2	**reaction progress 33,817**
initial solution	[55,508 O + 111,02 H + 6,0605 Na + 6,0605 Cl]
initial reactants	27,754 $CaSO_4$ (anhydrite)
final products	27,754 $CaSO_4 \cdot 2\ H_2O$ (gypsum) + 6,06051 NaCl (halite)
final solution	[–]

Table 5. **Evolution of brine chemistry during the polyhalitisation of anhydrite by a carnallite saturated (Q) solution. EQ3/6 modelling, 25°C, HMW database.**

Element Totals for the Aqueous Solution								
reaction	elements [mol]							
progress	S	O	Ca	Cl	H	K	Mg	Na
0	0,28313	56,642	0,000284	8,7264	111,02	0,6051	4,074	0,5392
1E-09	0,28313	56,642	0,000284	8,7264	111,02	0,6051	4,074	0,5392
3,1623E-09	0,28313	56,642	0,000284	8,7264	111,02	0,6051	4,074	0,5392
1E-08	0,28313	56,642	0,000284	8,7264	111,02	0,6051	4,074	0,5392
3,1623E-08	0,28313	56,642	0,000284	8,7264	111,02	0,6051	4,074	0,5392
1E-07	0,28313	56,642	0,0002841	8,7264	111,02	0,6051	4,074	0,5392
3,1623E-07	0,28313	56,642	0,0002843	8,7264	111,02	0,6051	4,074	0,5392
1E-06	0,28313	56,642	0,000285	8,7264	111,02	0,6051	4,074	0,5392
3,1623E-06	0,28313	56,642	0,0002871	8,7264	111,02	0,6051	4,074	0,5392
1E-05	0,28314	56,642	0,000294	8,7264	111,02	0,6051	4,074	0,5392
3,1623E-05	0,28316	56,642	0,0003156	8,7264	111,02	0,6051	4,074	0,5392
0,0001	0,28323	56,642	0,000384	8,7264	111,02	0,6051	4,074	0,5392
0,00031623	0,28344	56,643	0,0006002	8,7263	111,02	0,605	4,074	0,5392
0,0006086	0,28374	56,644	0,0008919	8,7262	111,02	0,605	4,074	0,5391
0,001	0,28335	56,643	0,0008943	8,7266	111,02	0,605	4,074	0,5391
0,0031623	0,28121	56,632	0,0009073	8,7286	111,01	0,6048	4,073	0,5391
0,01	0,27446	56,598	0,0009506	8,7349	111	0,6044	4,07	0,5391
0,031623	0,25316	56,491	0,0011103	8,7548	110,96	0,6029	4,059	0,5391
0,042679	0,2423	56,437	0,0012082	8,765	110,93	0,6021	4,054	0,5391
0,054026	0,23123	56,381	0,0013458	8,7659	110,91	0,5909	4,048	0,5399
0,1	0,18695	56,159	0,0021944	8,7659	110,82	0,5457	4,025	0,5399
0,1157	0,17231	56,086	0,0026427	8,7659	110,79	0,5307	4,018	0,5399

Table 6. **Evolution of reactants and products during the polyhalitisation of anhydrite by a carnallite saturated (Q) solution. EQ3/6 modelling, 25°C, HMW database.**

Reactant and Product Summary					
reaction progress	reactants [mol] anhydrite	products [mol] sylvite	halite	polyhalite	anhydrite
0	0	0,03878	0,0007017	0	0
1E-09	1E-09	0,03878	0,0007017	0	0
3,1623E-09	3,1623E-09	0,03878	0,0007017	0	0
1E-08	1E-08	0,03878	0,0007017	0	0
3,1623E-08	3,1623E-08	0,03878	0,0007017	0	0
1E-07	1E-07	0,03878	0,0007017	0	0
3,1623E-07	3,1623E-07	0,03878	0,0007017	0	0
1E-06	1E-06	0,03878	0,0007017	0	0
3,1623E-06	3,1623E-06	0,03878	0,0007024	0	0
1E-05	1E-05	0,03878	0,0007039	0	0
3,1623E-05	3,1623E-05	0,03878	0,0007088	0	0
0,0001	0,0001	0,03878	0,0007241	0	0
0,00031623	0,00031623	0,03879	0,0007726	0	0
0,0006086	0,0006086	0,03881	0,0008381	3.043E-07	0
0,001	0,001	0,03845	0,0008381	0.0001948	0
0,0031623	0,0031623	0,03645	0,0008385	0.0012694	0
0,01	0,01	0,03013	0,0008403	0,0046667	0
0,031623	0,031623	0,01019	0,0008505	0,015398	0
0,042679	0,042679	0	0,0008591	0,020877	0
0,054026	0,054026	0	0	0,026482	0
0,1	0,1	0	0	0,049045	0
0,1157	0,1157	0	0	0,056589	0,000166

Table 7. **Evolution of the mineral saturation states during the polyhalitisation of anhydrite by a carnallite saturated (Q) solution. EQ3/6 modelling, 25°C, HMW database.**

Summary of Pure Mineral Saturation States				
reaction progress	affinities [kcal] anhydrite	halite	polyhalite	sylvite
0	-1,04	0	-1,361	0
1E-09	-1,04	0	-1,361	0
3,1623E-09	-1,04	0	-1,361	0
1E-08	-1,04	0	-1,361	0
3,1623E-08	-1,04	0	-1,361	0
1E-07	-1,04	0	-1,361	0
3,1623E-07	-1,04	0	-1,36	0
1E-06	-1,038	0	-1,357	0
3,1623E-06	-1,034	0	-1,348	0
1E-05	-1,02	0	-1,32	0
3,1623E-05	-0,978	0	-1,236	0
0,0001	-0,862	0	-1,003	0
0,00031623	-0,596	0	-0,472	0
0,0006086	-0,361	0	0	0
0,001	-0,36	0	0	0
0,0031623	-0,356	0	0	0
0,01	-0,342	0	0	0
0,031623	-0,294	0	0	0
0,042679	-0,268	0	0	0
0,054026	-0,231	0	0	-0,011
0,1	-0,063	-0,004	0	-0,057
0,1157	0	-0,006	0	-0,073

Table 8. **Quantitative reaction equations during the polyhalitisation of anhydrite by a carnallite saturated (Q) solution. EQ3/6 modelling, 25°C, HMW database.**

Reaction summary	reaction progress 0,1157
initial solution	[56,642 O + 111,02 H + 0,539 Na + 0,605 K + 2,8E-04 Ca + 4,074 Mg + 8,726 Cl + 0,283 S]
initial reactants	0,11553 $CaSO_4$ (anhydrite)
final products	0,05659 $K_2MgCa_2(SO_4)_4 \cdot 2\ H_2O$ (polyhalite)
final solution	[56,086 O + 110,79 H + 0,540 Na + 0,531 K + 2,64E-03 Ca + 4,018 Mg + 8,766 Cl + 0,172 S]
Detailed reaction	**reaction progress 0,1157**
initial solution	[56,642 O + 111,02 H + 0,539 Na + 0,605 K + 2,8E-04 Ca + 4,074 Mg + 8,726 Cl + 0,283 S]
initial reactants	0,1157 $CaSO_4$ (anhydrite)
in between products = reactants	0,000859 NaCl (halite) + 0,010187 KCl (sylvite)
final products	0,00017 $CaSO_4$ (anhydrite) + 0,05659 $K_2MgCa_2(SO_4)_4 \cdot 2\ H_2O$ (polyhalite)
final solution	[56,086 O + 110.79 H + 0,540 Na + 0,531 K + 2,64E-03 Ca + 4,018 Mg + 8,766 Cl + 0,172 S]

EVALUATION OF LONG-TERM GEOLOGICAL AND CLIMATE CHANGES IN THE SPANISH PROGRAM

Carlos del Olmo Alonso

ENRESA. Empresa Nacional de Residuos Radiactivos
Emilio Vargas, 7 - 28043 MADRID, SPAIN

ABSTRACT.

Assessment of the long-term evolution of deep repositories involves evaluation of the interactions between the components of the disposal system following different evolution scenarios constructed to fix the boundary conditions of the multibarrier system at different spatial and time scales.

As a dynamic system the characterization of the geological barrier is mastered by its past evolution, which will guide us, following the Lyell's "actualism" paradigm, in the construction of future evolution scenarios.

Siting and characterization of a disposal site must consider first order processes on a global scale (climate, plate tectonics) that can modify processes or events at regional scale (change in climatic parameters, change in geodynamic processes) that will influence processes and events at areal and local scale (state of stress/seismicity, hydrology, hydrogeological system, erosion, ...). In view of this, ENRESA has established a major research programme, in cooperation with the Geological Survey of Spain, also involving several academic institutions and research centres, whose main purpose, orientation and findings are outlined.

1. INTRODUCTION.

Assessments of the long-term evolution of a deep repository for long-lived, high level radioactive waste, involve evaluation of the interactions between the components of the disposal system following different evolution scenarios through an in-depth understanding of the system. Construction of the different evolution scenarios and probabilities of occurrence is not a simple task but it is essential for fixing the boundary conditions of the multibarrier system at different spatial and time scales.

As a dynamic system, the characteristics of the geological barrier are mastered by its past evolution, which will guide us, following the principles of Sir Charles Lyell "actualism" paradigm: processes observed today are operated in the same way as in the past and will continue to be operated in the same way in the future (1) in the construction of the "non altered" or central evolution scenario. So siting and characterization studies of a potential disposal site must:

- describe the initial state of the site, defining its characteristics and past history, and
- construct an inventory of processes and events that can influence its evolution.

This inventory must:

- define the relative importance of the different phenomena.
- define the interactions between them, and
- define its time and spatial dependence.

Siting and characterization programmes must provide a hierarchization of the processes, considering their spatial and temporal scale, (global, regional, areal, local ...) and establish cause-effect relations to:

- First, construct a conceptual (qualitative) description of the site characteristics, its evolution and the components and characteristics of the disposal system.
- Second, through the characterization and associated experimental research programmes, quantify the parameters that control the system behaviour and its associated uncertainty.

This phenomenological approach must be iterative and phased to optimize the considerable scientific and economical resources necessary to solve the problems involved.

As an example, siting and characterization of a disposal site must start by defining the 1st order processes that, at global and geological scales (climate, plate tectonics) will control 2nd order processes or events acting at a regional scale (glaciation, change in climatic parameters, state of stress, faulting, elevation, subsidence, volcanism,...) that will influence processes and events at areal and local scale (hydrology, hydrogeological system, geochemistry, sea level, erosion, seismicity, ...).

Since 1986 ENRESA has established a research program, in cooperation with the Spanish Geological Survey (ITGE) and several Spanish and foreign institutions (Universities, CSIC, BRGM, ...) oriented towards geodynamic and climate research, whose main purpose, orientation and findings are outlined.

2. STUDIES IN GEODYNAMIC RESEARCH.

Among the main R&D activities developed to support the site evaluation and characterization programme (geological disposal) we must mention the following projects:

2.1 Basic Studies.

- Neotectonic, Seismotectonic and Seismic Risk Evaluation in Spain.

  ENRESA promoted a joint work between the Spanish Geological and Geographical Survey (ITGE, IGN) and several universities to carry out this research. The results were published in 1992 in the form of a Neotectonic and Seismotectonic Map of Spain at a scale 1:1.000.000 as a summary of the studies performed at 1:200.000 scale throughout Spain.

One of the main conclusions of this Project was the construction of a geodynamic model (J. Baena et al., 92) (2) for the Iberian Peninsula based on differential extrusion between the African and Eurasian plates. (Fig. 1).

CLOSING AFRICAN and EURASIAN PLATES (NNE-SSW)

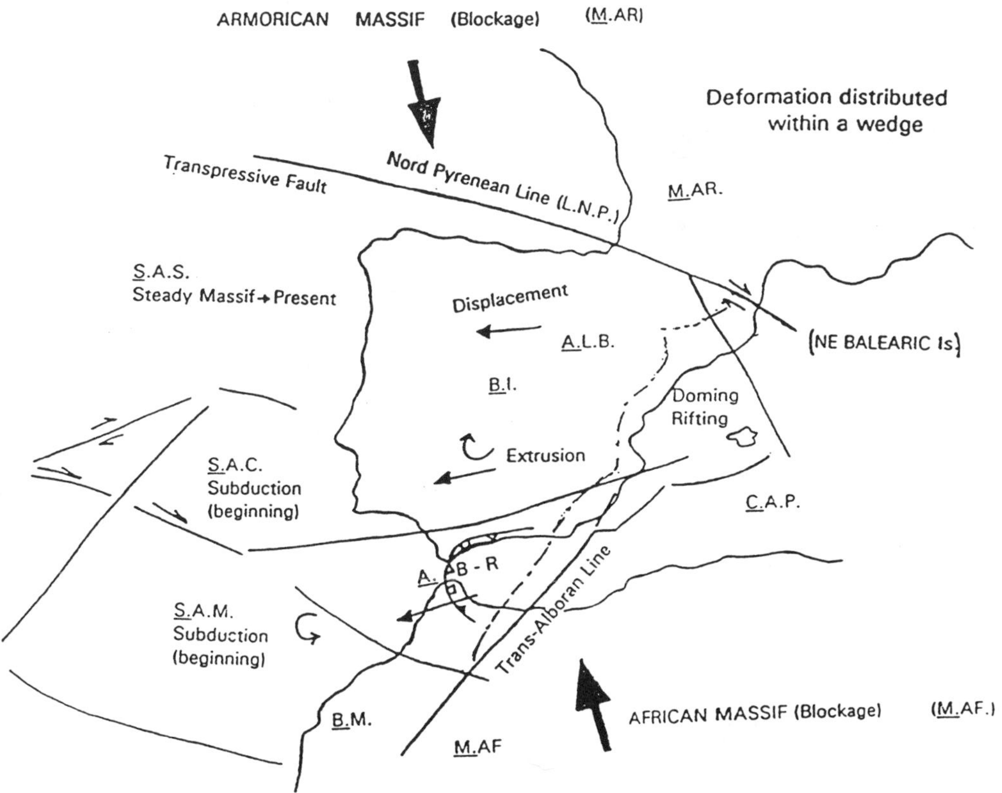

Fig. 1. Geodynamic model for the Iberian Peninsula.

- Review of Past Earthquakes.

ENRESA has co-sponsored a collaboration between the Spanish Nuclear Safety Council (CSN) and the IGN for this task. It's main goal was a re-assessment of the epicentral intensity of key historic earthquakes and the elaboration, from these data, of new isosists maps.

National archives with sources of information since 14th Century, were studied in detail, but the information obtained does not modify the relevant characteristics of the National Geographical Institute "Earthquakes Catalogue".

• Review of fragile fractures and Regional State of Stress Characterization in the NW part of the Iberian Massif.

The main objective of this study, carried out using remote sensing techniques, by the Geodynamic Department of the Complutense University of Madrid was the characterization of the fracturing and regional state of stresses, and its evolution, placing a special emphasis on the determination of the current state of stress in the area of study, the Hesperian Massif, because this is an essential factor in the analysis of fracture permeability.

The work was done by analysis of the alignments identified from the Landsat imagery, as fractures or dykes, in a region of around 200.000 Km^2 extended from the granitic outcrops of the "Sistema Central" to the western limits of the Iberian Massif.

The methodology used in the study is presented in Fig. 2. The current state of stress defined by the populational analysis of alignments has been compared with field data arising from overcoring, population analysis of recent faults and focal mechanisms. Those data slightly modify the regional directions from alignments analysis. Final conclusion about current regional direction of the main stresses in Spain are presented in Fig 3.

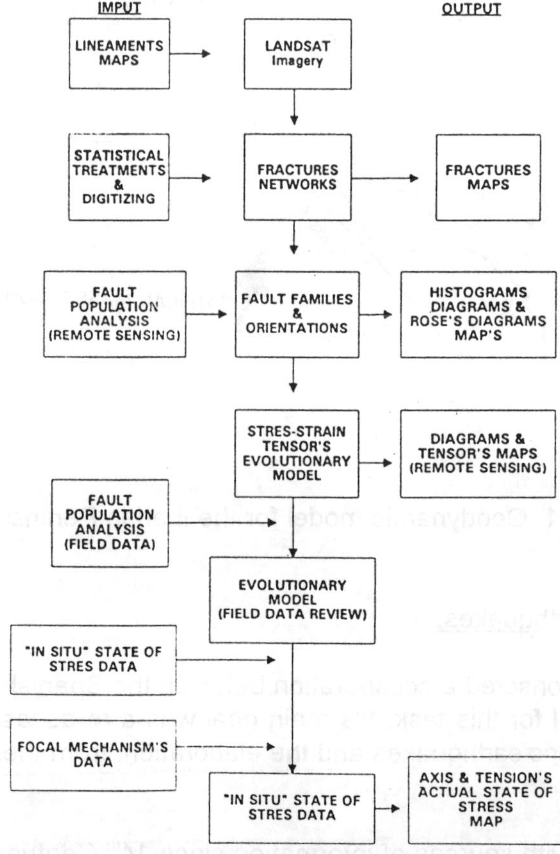

Fig. 2. Review of state of stress and fragile fractures in the Hesperian Massif. Methodology.

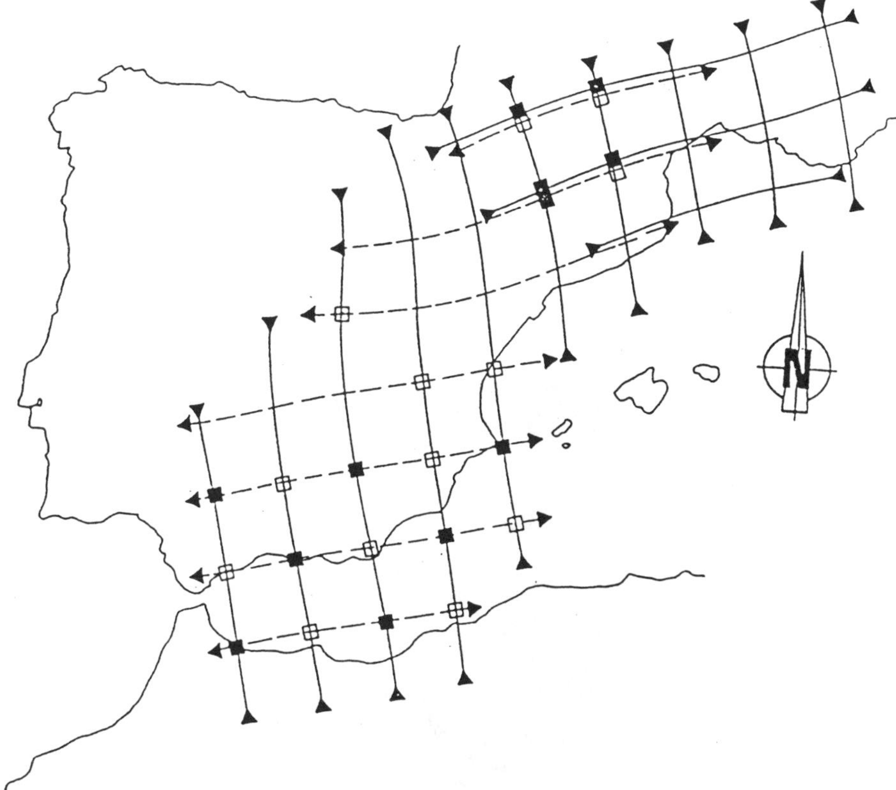

Fig. 3. Regional direction of the main stresses in Spain.

- Seismic Studies.

Since the late 80's a very major effort has been devoted to studying the upper mantle structure and Iberian lithosphere in Spain using high resolution seismic methods, with a view to increase the knowledge of these parts of the crust and to establishing correlations between crustal structure and distribution of seismicity in tectonically active areas. The results of two of the main Programmes, ESCI and ILIHA, were recently available (Fig. 4).

ESCI Programme.

The ESCI Programme (acronym in Spanish for Seismic Studies of the Iberian Crust) was initiated in 1990 by the Spanish National R+D Plan and partially funded by the CEC.

This Programme has established deep seismic reflection research in key geodynamic areas of the Iberian Peninsula:

- The North-West area of the Hercynian Massif.
- The Pyrenees-Valencia trough.
- The Betics.

The efforts of most Spanish scientists in these three projects working in these earth science projects was integrated, and the main objective was to study the deep crustal structure of the Iberian Peninsula by seismic near-vertical reflection methods, focusing on the analysis of:

- Nature and structure of the crust.
- Development of collisional structures.

- Response to extensional stresses in regions of recently thinned crust.
- Correlation between crustal structure and distribution of seismicity in tectonically active areas.

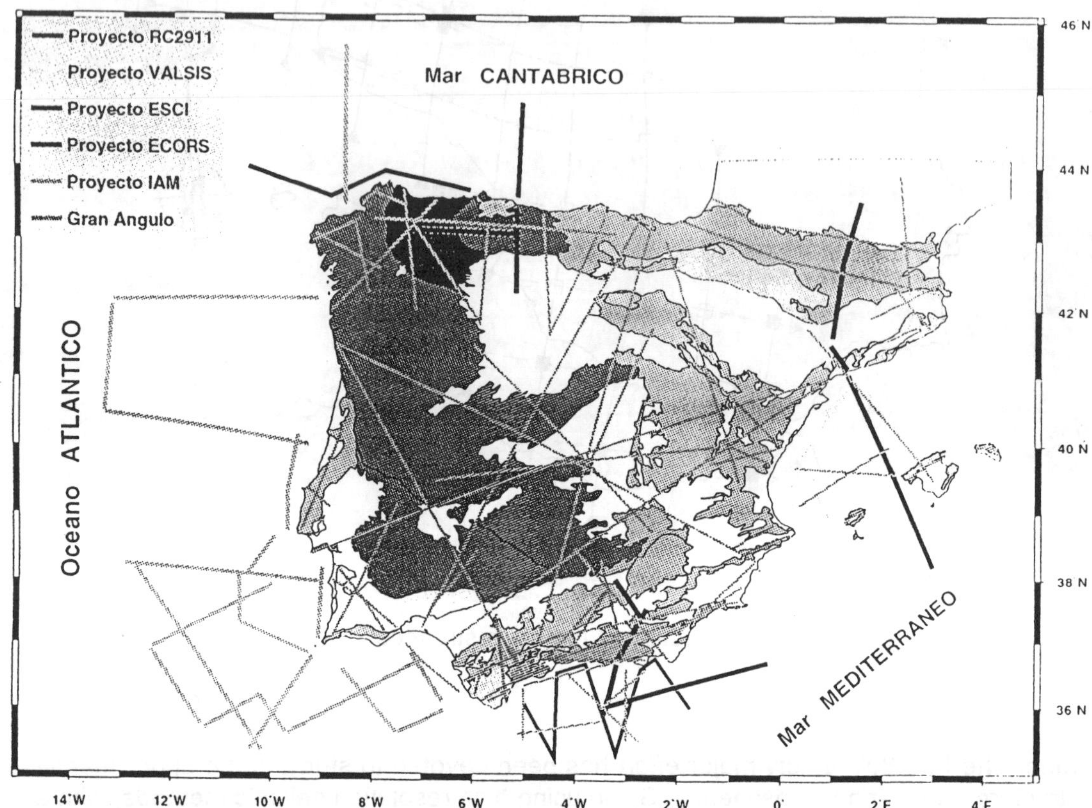

Fig. 4. Wide angle and deep seismic reflexion profiles in the Iberian Peninsula.

<u>ILIHA Programme.</u>

The ILIHA Programme (Iberian Lithosphere Heterogeneously and Anisotropy) has used advantage of the technology and methodologies developed by the participants involved in the EGT (European Science Foundation "GEOTRAVERSE" Project) and was co-funded by the CEE in 1988.

The aim of the Project was to increase the knowledge of the Iberian lithosphere and upper mantle structure with as high a resolution as possible by means of seismological studies. The ILIHA Project contributes to the EGT whose main aim is to investigate the nature, dynamics and evolution of the continental lithosphere underlying Europe.

The upper mantle structure, over a depth range from 30 to 100 Km has been investigated by two methods.

- analysis of earthquake surface and body waves recorded by the broadband stations of the permanent Spanish and Portuguese observatories and the portable NARS and

- a large scale Deep Seismic Sounding experiment in which six profiles sampled the lower lithosphere over a central region in the Hercynian core of Iberia.

The data were collected at 14 NARS stations and 140 mobile stations concentrated along 8 profiles under the framework of the Project. Most of these were in operation for almost one year. The installation of the NARS array has recorded body and surface waves generated by distant and local earthquakes and a large amount of digital data is available.

The seismograms analyzed by studying fundamental mode Rayleigh wave dispersion have defined the lateral inconsistencies in the earth's upper mantle, and a new regional model for the deep structure of Iberia down to a depth of 200 Km was proposed.

- Review of surface & subsurface geology of Spain.

ENRESA, in cooperation with the Spanish Geological Survey (ITGE), promoted a review and synthesis of the documentation obtained by several companies and research institutions in hydrocarbon exploration, mining exploration (coal, salts, metals, ...) and underground water resources exploration in Spain. Most of this information was not confidential and was available in several files of the Central and Autonomical Regions Administrations.

The work started in 1989 by analysing the data of a synthesis made by REPSOL (Spanish Hydrocarbon Exploration Company) in 1974 in their permits and in the permits of other associated companies (Coparex, Texaco-Chevron, Philips, Esso and others), this synthesis also includes information obtained through exchange with other companies (Campsa, Cepsa, Shell, etc.).

At the same time, ENRESA performed a review of surface geological data and data from boreholes as a result of Uranium exploration (ENUSA & CIEMAT). This review also includes a reprocessing and reinterpretation of gravimetric, aeromagnetic and radiometric data.

Finally, in the review other documents were integrated: data coming from the National Hydrological Plan, from mineral exploration (Navarra and Ebro Basin Potash exploration) and other bibliographical sources, mainly from research and educational centres (Universities, CSIC, ICONA, ...).

A synthesis of the data available in the sedimentary basins was publicly presented by the Spanish Geological Survey in 1991 (3). The results of the reinterpretations of the gravimetric, aeromagnetic and radiometric data were presented in 1992 (4).

2.2 Applied studies.

- <u>Seismic risk in fault segments.</u>

  This activity, currently under progress, is being developed by the Geodynamic Department of Universidad Complutense de Madrid. Its main objective is to refine the methods used to define the seismic risk and long-term stability in areas of low to moderate seismicity, as is the case for most parts of the Spanish territory that could be used for geological disposal of radioactive waste.

  After a review of different fault types, a structure with more than 500 Km in length, named "Plasencia-Fault", has been chosen for methodological purposes. This fault, mainly a tectonic graben, has a set of associated dykes and faults and has been reactivated several times in different geological periods. One of the objectives of the work is to distinguish between paleotectonic and neotectonic movements, particularly since the Cenozoic times.

  At the same time as field studies took place, a review of the seismic data in a band of 60 Km at both sides of the fault was performed. Both intensity and magnitude data were reviewed. Seismic episodes spatial distribution shows how the seismicity is in the vicinity of the intersections points of other faults associated with the Plasencia Fault and never in the Plasencia Fault itself.

  Geophysical, morphological and microseismicity studies are planned to help us to build the characterization methods and skills needed to deal with this kind of long-term geodynamic process.

3. STUDIES IN SUPPORT OF CLIMATE AND ENVIRONMENTAL EVOLUTION.

3.1 Objectives and scope.

The work described below is part of a CEC Project, FI2W-CT91-0075, entitled "Paleoclimatological Revision of Climate Evolution and Environment in Western Mediterranean Regions" currently under way and jointly sponsored by ENRESA. This Project was initiated as a result of the need to evaluate the impact that the future evolution of the environment might have on the disposal of high-level radioactive wastes on Spain. The Project, which consists of a geoprospective study, has two main objectives:

- To obtain information on climatic evolution and environmental changes on the Iberian Peninsula and surrounding Mediterranean countries during the Quaternary.
- Paleoenvironmental reconstruction of a site on the Peninsula over the last 100,000 years, and construction of future evolution scenarios.

The organizations participating with ENRESA in this Project are Instituto Tecnológico Geominero de España (ITGE), Consejo Superior de Investigaciones Científicas (CSIC), Escuela Técnica Superior de Ingenieros de Minas de Madrid (ETSIMM) and Bureau de Recherches Géologiques et Minières (BRGM).

Paleoclimatic and Paleoenvironmental research is developed on two timescales: the last two million years and the last thousand years. Three research projects have been initiated to improve on current knowledge of the environment in Spain during the Quaternary:

a) Research into Quaternary paleoenvironmental evolution of a sector of the River Tajo Valley, carried out by ITGE in collaboration with CSIC.

b) Paleoclimatic reconstruction from the middle Pleistocene on the basis of geochronological and isotopic analysis of Spanish carbonate deposits, carried out by ITGE in collaboration with ETSIMM.

c) Climatic reconstruction of the Spanish mainland over the last thousand years on the basis of dendrochronological series, carried out by ITGE and CSIC.

The work carried out by BRGM basically consists of applying in-house geoprospective models and software for paleoenvironmental reconstruction and construction of future evolution scenarios. The tasks performed are as follows:

a) Western Europe and the Iberian Peninsula from -120,000 years to the present (computerized construction maps).

b) Modelling of one hundred thousand years of geological evolution at the Jarama paleosite (Spain).

c) Simulation of the future evolution of the Iberian Peninsula over a timescale of 100,000 years.

3.2 The Quaternary on the Iberian Peninsula: an assessment.

The Iberian Peninsula is made up of a wide variety of geotectonic units (5) (Figure 5), consisting of a relatively stable Central part, dominated by plateau areas, and two heavily deformed Alpine Chains at its Northern and Southern edges (the Pyrenean Chain, with its western prolongation the Vascogothic Chain, and the Southern Spanish or Betic Chain, which stretches to the Balearic Islands). The present geodynamic framework of the Iberian Peninsula is that of a situation inherited from the alpine structure of the region. In geological terms, this is a response to the relative movements between the African and Eurasian plates, and to the role played by the Iberian plate during development of the Atlantic Ocean. Although the stress field caused by the approximation of Europe and Africa has decreased considerably, it continues to act today. This compressive regime has persisted more in the southern area of the Iberian Peninsula, close to the zone of contact with the African Plate. In the rest of the territory the neotectonic manifestations are almost entirely distensive.

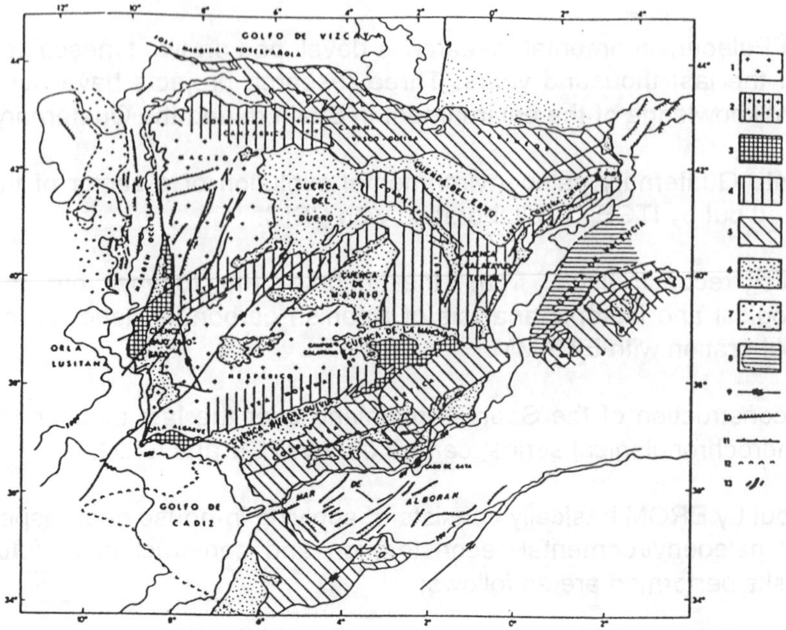

1) Hercynian basement in the Hesperian Massif. 2) Zones elevated by compressional tectonism in the Hesperian Massif. 3) Subhorizontal mesozoic in borders of Hesperian Massif and mesozoic with moderate deformation in Lusitanian and Algarvian borders. 4) Intermediated Folded Chains. 5) Alpine Cordilleras. 6) Tertiary basins. 7) Graben with very thick Early Cretaceous sediments in western continental margin. 8) Zone with eroded Mesozoic in the central part of Valencia Trough. 9) Fault showing recent transcurrent-like tectonic activity. 10) Fault showing recent normal tectonic activity; submarine morphological escarpment. 11) Fault showing recent reverse tectonic activity. 12) Front of the Olistostrome and slipped units. 13) Volcanic areas.

Fig. 5. Tectonic sketch of Iberian Peninsula (Capote and Vicente, 1989).

Quaternary evolution on the Iberian Peninsula is subject more to a dissection model than to a model of sediment accumulation. Sediments are generally not very thick, with the exception of those found in the Guadix-Baza Basin. Dating or establishing stratigraphic correlations in these deposits is not easy. Practically all the Quaternary deposits are continental, with predominance of fluvial terraces, alluvial plains, glacis, Piedmont and alluvial cone deposits. Coarse fragmentary deposits prevail, indicating in the majority of cases the transport of bed loads, which makes the accumulation and preservation of flora and fauna remains unlikely, and complicates correlation or dating techniques. Additionally, the scarcity of deposit sequences covering long periods of time during the Quaternary, along with the relative scarcity of fossil remains, complicates enormously the establishment of accurate chronologies, and even the acquisition of significant paleoenvironmental data. This shortcoming is especially evident on the Iberian Massif, with the exception of the karstic fill of the Sierra de Atapuerca (Burgos), covering the Middle Pleistocene and part of the Lower, where a multidisciplinary team has been working for more than 10 years. Other well known sites, such as the Torralba and Ambrona (Soria) sequences, and the Redueña and Aridos deposits in the Madrid basin, cover a much shorter period of time than Atapuerca.

The degree of knowledge of the Quaternary on the Iberian Peninsula is high as regards the mapping of deposits and land forms. Important gaps still exist, however, with respect to the chronology and paleoenvironmental and paleoclimatic significance of these deposits and forms. Although an important volume of data is available in different publications, unpublished works, etc., overall knowledge of the Quaternary environment is clearly insufficient. The information available refers to partial aspects of the environment, and belongs to different fields of scientific research. In particular, knowledge of the climate during this period consists generally of qualitative data (trends), in most cases with rather inaccurate relative chronologies.

- **Paleoclimatic and environmental study of Quaternary deposits in the Tajo Valley.**

Erosion was the dominant process in the centre of the Iberian Peninsula during the Quaternary. It has formed a complex system of terraces and glacis in the major river valleys of the Iberian Peninsula. The geomorphology of these systems is relatively well known, but important gaps still exist regarding their chronology and their paleoenvironmental significance. Of all the major Spanish valleys, the Tajo Valley , with its well preserved terraces containing fossils, archaeological sites and buried soils, was considered a suitable site for obtaining climatic and environmental information. At the scale of the Iberian Peninsula, the results obtained in the Tajo Valley will be of great use because of the evolutionary similarity between this valley and other major valleys in the Peninsula.

The ultimate objective of this task is to establish the relationship between the terraces along the Tajo Valley and the isotopic states of oxygen. This could later be used to establish paleoclimatic correlations with the rest of Europe.

Several fossils and a polarity change found in these deposits have been used to date some terraces and to establish correlations between them in two different areas. Some fossils found, mainly mammals and pollen, have provided useful information about the predominant environmental conditions in the Tajo Valley during the Quaternary.

- **Study and dating of travertines in Spain as a paleoclimatic and paleoenvironmental index.**

The objective of this task is to investigate the possibilities of using quaternary continental carbonate deposits as paleoclimatic and paleoenvironmental indicators. These deposits were selected because they are very abundant and widely distributed throughout the Iberian Peninsula. However, the type of environmental information they provide differs from one type of carbonate deposit to the other. For instance, the isotopic composition of travertine depends on several factors that are difficult to control and which may introduce considerable doubts as regards the paleotemperatures deduced from them. In addition, travertines are not usually preserved during cold periods when fluvial incision is prevalent. However, they often contain fossil remains, such as pollen, gasteropods and vertebrates that can be useful climate indicators. On the other hand, the paleoclimatic information obtained from speleothems is less problematic because their isotopic composition is affected by fewer variables.

The research was centred on five field sites that meet most of the following criteria: well preserved deposits of different origin with long stratigraphic records, fossil remains and situated in different climatic areas throughout the Peninsula.

The information collected at the sites with travertine deposits has made it possible to date all the travertine terraces found and to reconstruct the paleoenvironmental evolution of the sites. This will be used to calculate incision rates and erosion. Pollen and isotope studies carried out in the sediments of a small lake in Northern Spain suggest that stable isotopes in travertines could be used as paleoclimatic indicators. Isotope analyses in speleothems indicate progressive cooling since the Middle Pleistocene, with short intervals of higher temperatures.

- **Climatic reconstruction of the last thousand years in Spain on the basis of dendrochronological series.**

  Thirty-seven chronologies (i.e. each tree ring has been matched with the corresponding year of formation) have been obtained. They have been grouped in seven zones to increase the reliability of climatic retropredictions. (Fig. 6).

  To reconstruct past climatic variables it is necessary, in addition to establishing chronologies for each area, to collect climatic information from existing weather stations. This objective was also achieved during 1994. Only those weather stations with rainfall and temperature records dating back for over 40 years have been selected.

  Several climatic variables have already been reconstructed in seven weather stations, whose location can be seen in Figure 6. So far, sequences up to 300 years have been reconstructed.

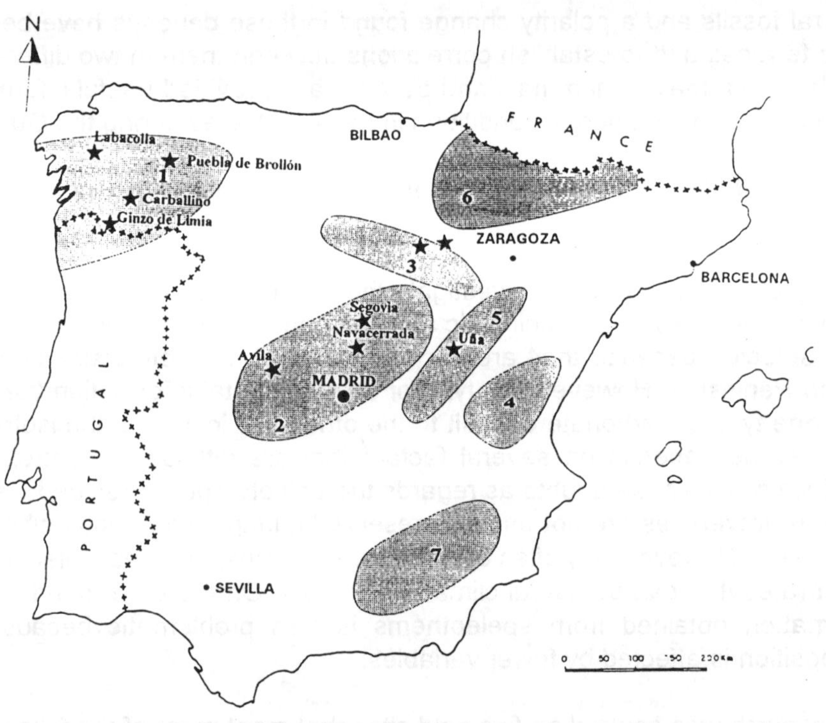

* Weather stations with climatic reconstruction.

Fig. 6. Climatic areas for tree-ring studies.

- **Paleoenvironmental reconstruction and construction of future evolution scenarios.**

  The first objective, the reconstitution of a site representative of conditions in Spain, during the last glaciation has been completed. The reconstitution has been carried out with the code "Prospect" developed by BRGM. The work has involved the identification of all the relevant geological processes that have acted upon this site over the last 120.000 years and, whenever possible, their quantification. (6) (7).

The site chosen is located in the Tajo Basin, near the Jarama River. The basin was formed during the late Cretaceous and filled up with fluvial, lacustre and palustre sediments during the Tertiary. The Quaternary began with the deposition of Piedmont deposits, known as Raña, followed by fluvial dissection, which created a system of terraces, glacis and alluvial fans typical of most Spanish valleys. Some of the rivers at the site seem to be controlled by a fault system that may have been active during the Quaternary. A hypothetical repository is lies at a depth of 500 m below the surface of the ground in the Tertiary sediments.

Figure 7 shows schematically the conceptual model developed for the site. That is, the processes considered to have affected the site and the links established between them.

The simulated causes of site evolution are sea level fluctuations due to freezing and melting of glaciers worldwide, annual mean global temperature, regional vertical movements and differential vertical displacement caused by a hypothetical fault. The parameters that quantify these processes are represented by the uppermost nodes, or entry nodes, in Figure 7. Their magnitude and rate of change are input to the model at the beginning of the simulation.

The rest of the nodes in Figure 7, are no-entry nodes, and represent parameters that are calculated by the code. Their values at any given time depend on the entry node values and the links established between the nodes.

The causal links established between the parameters depend on whether the parameter is a variable (i.e. a site characteristic such as temperature) or a phenomenon (i.e. primary cause of evolution with time). Deriv represents the impact of a phenomenon on a variable. CDeriv represents the impact of a phenomenon on a variable under certain conditions. M+ or M- is used when a change in a variable causes proportional changes in another variable. CM+ is similar to M+ but with certain conditions. CDcl indicates the link established if a variable reaches a critical value that may trigger off a phenomenon. IDecl is similar to CDecl, but a value is assigned to the phenomenon. Equa is used when the value of a variable is the result of a mathematical relation between many variables.

During simulation, entry node values and their evolution over the last 120.000 years are propagated through the conceptual scheme described above. The model is calibrated when the results obtained compare well with present field conditions.

The best simulation was achieved when vertical movements were introduced in the entry nodes. This made it possible to simulate tilt blocks which are believed to have caused the migration of the Jarama River. The results show that over the last 120.000 years the Jarama has eroded 15 m of river bed deposits and created 3 terraces of 1 to 2 metres in thickness. The temperature at the simulated site would not allow permafrost formation.

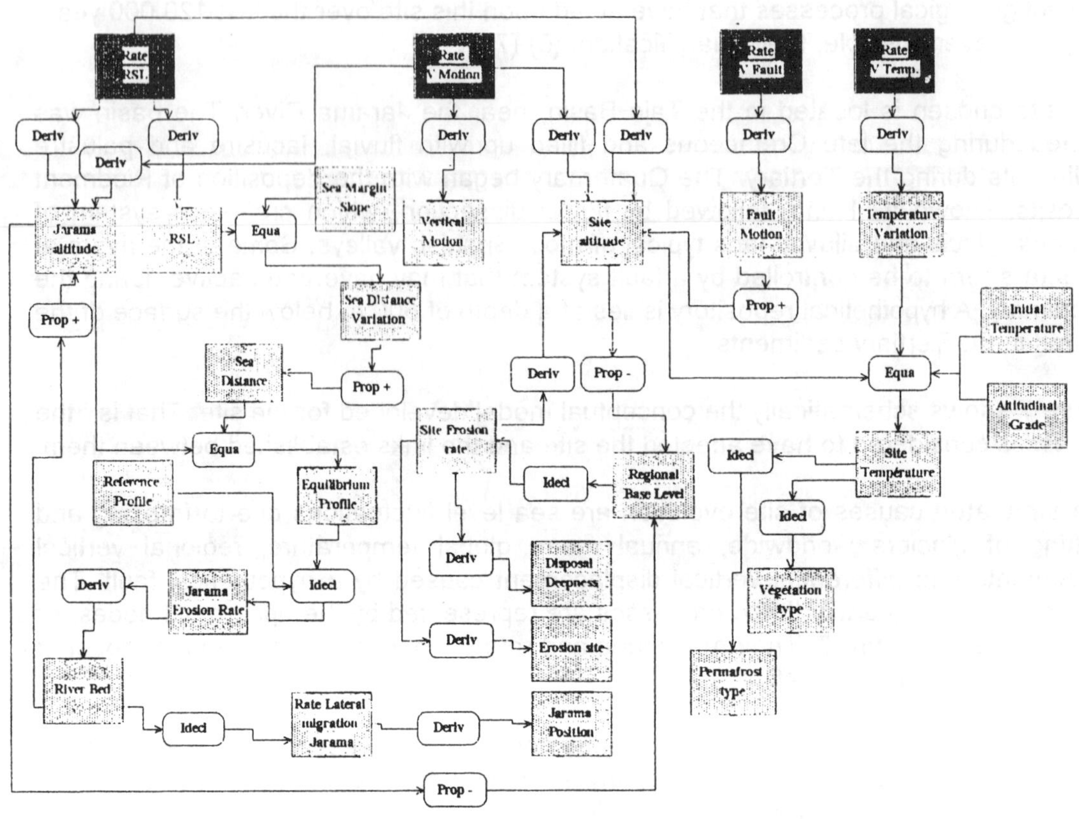

Fig. 7. Processes and relationships considered in the paleoenvironmental reconstitution of the Jarama River Site.

In addition, work has progressed in the construction of future evolution scenarios at the paleosite for the next 100,000 years. It is assumed that the geological and climatological conditions during the next 100,000 years will be comparable in amplitude and sequence to those experienced during the most recent Quaternary. The orbital curves presented by Imbrie and Imbrie (1980) (8) and ACLIN1 have been used to simulate future temperature and sea level changes. The best correlation found between these curves and the paleoclimatic records indicate that it is better to use the ACLIN1 curve for temperature variations and the Imbrie curve for sea level changes. (Figure 8). It is also assumed that the stress field at the paleosite will not change greatly during the next 100,000 years, since the site is not expected to be subjected to glacial isostasy or major changes due to plate tectonics. The rest of the information needed for the future evolution of the paleosite is the same as for the calibration of the paleosite. Some preliminary results and schematic reconstitutions are presented in Figure 9.

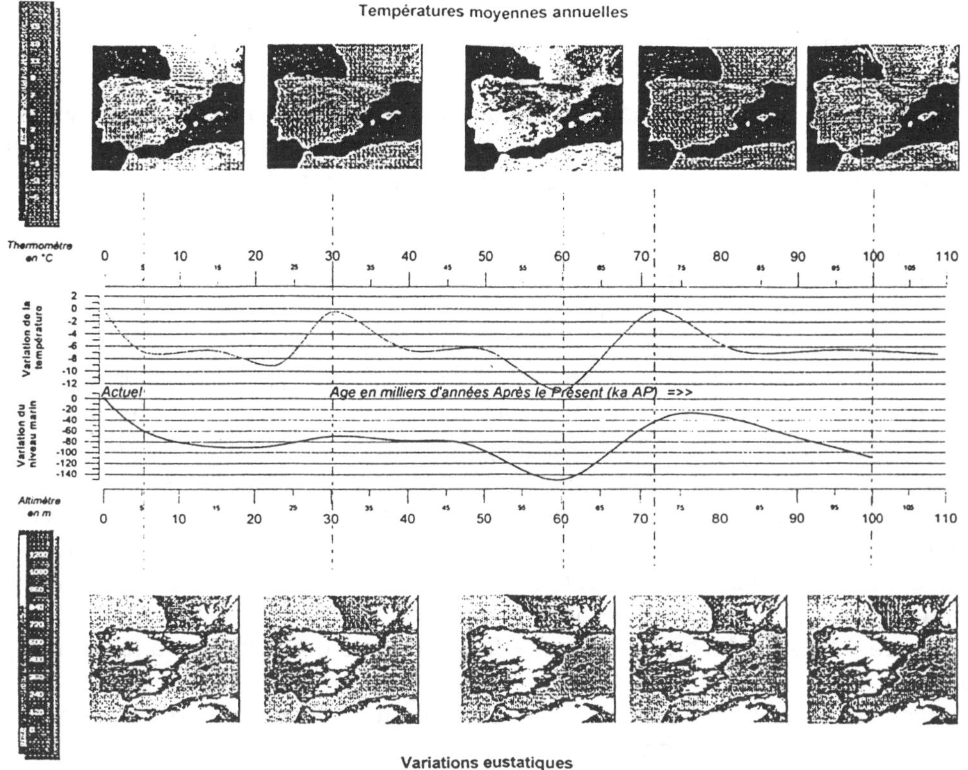

Fig. 8. Mean temperature and sea level changes considered for the next 100,000 years.

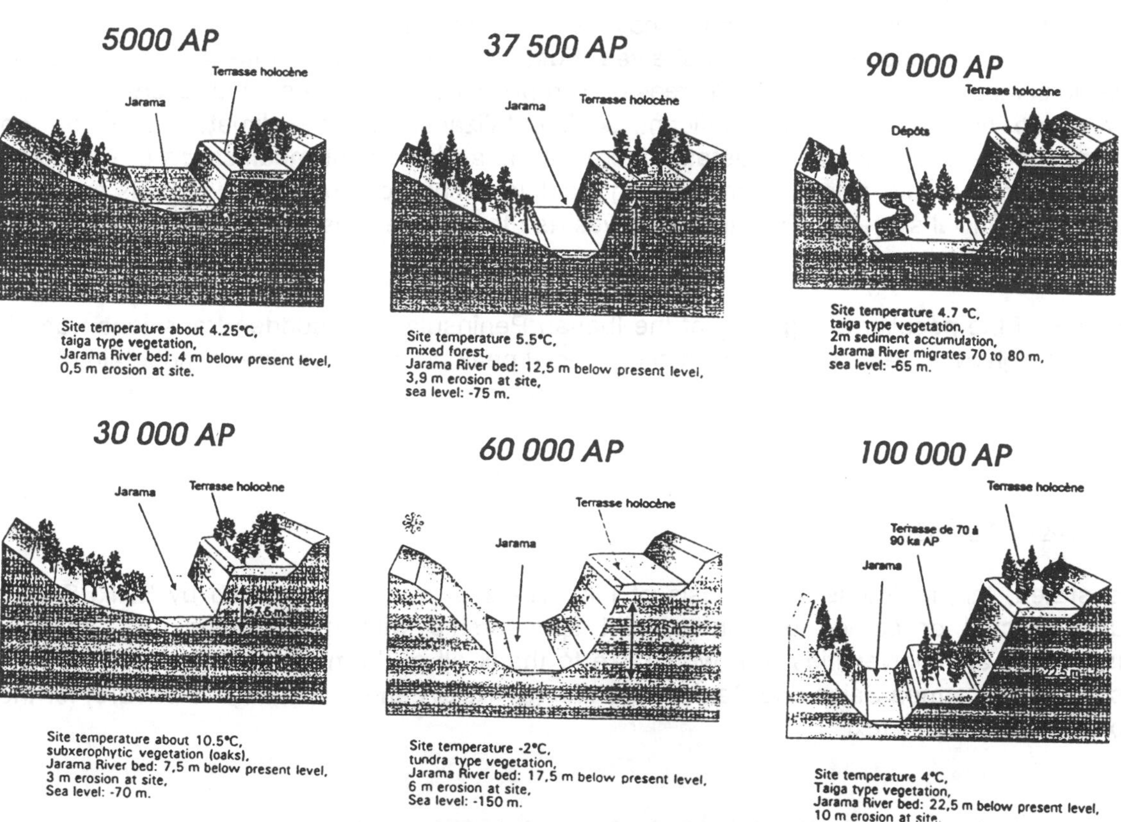

Fig. 9. Site evolution for the next 100,000 years.

4. RESULTS. IMPLICATIONS FOR THE FUTURE

The ultimate objective of the work being carried out is to develop a geoprospective-surveying research methodology supporting analysis of the safety of radioactive waste disposal facilities in deep geological formations within the framework of the Spanish geographical environment.

The results of these paleoenvironmental reconstructions will contribute to filling the gaps in our present knowledge of the climatic and geodynamic evolution of the Iberian Quaternary, especially in the interior of the Peninsula. The aim is to gain insight into the relationship between the climatic changes that occurred during the Quaternary and the processes of external geodynamics (alteration, erosion, transport, sedimentation, etc.) that have led to the generation of deposits and forms. This will serve as a basis for the application of geoprospective techniques aimed at predicting the future evolution of the environment and geodynamic processes over a timescale of 100,000 years. The objective is to draw up coherent, realistic scenarios of the future evolution of the Iberian Peninsula, allowing the consequences of this evolution to be analyzed. The Iberian environment, shows certain differentiating characteristics with respect to other regions of the Centre and North of Europe. These peculiarities are due to the geographic position of the Peninsula, in the South of the continent between the Atlantic and the Mediterranean, to its orography and to the morphology of its continental shelf. As a result, there is currently an important diversity of climates (mountain, moist oceanic and Mediterranean), the Mediterranean type prevailing over most of Spain with varying degrees of aridness.

Spain is located outside the area which was disturbed by the glacial isostasis that arose as a result of northern inlandsis during the colder periods of the Quaternary. Nevertheless, the repercussions of eustatic variations may be important. The Iberian coastline shows very narrow, strongly sloped continental shelves. Eustatic variations affect the coastline relatively little, in comparison to more northern areas of Europe. This fact suggests that decreases in sea levels do not give rise to significant continentalization of the climate. However, one consequence of these glacio-eustatic decreases is an important modification of equilibrium profiles in rivers, this giving rise to a strong reactivation of erosion in the river beds and to the incision of fluvial systems. This, in turn, gives rise to surface erosion, due to the reduction of local base levels.

In view of the peculiar topography of the Iberian Peninsula, with sudden transitions from the coast to high average altitudes in the interior, glacial peak periods give rise to mountain glaciers in the major chains and to the establishment of a permafrost which may affect large areas of the interior, in spite of its southern location. (9).

Ackowledgements.

As stated, this review is a synthesis of nine years of studies promoted by ENRESA and subcontracted or carried out under joint sponsorship with many companies, research institutions and Universities. We would like to thank all of them for their collaboration, and especially to Carmen Bajos and Daniel Barettino (ENRESA and ITGE Project leaders) for their contributions to this text.

REFERENCES.

(1) LYELL, Sir Charles, 1854. "Principles of Geology". 9th Edn. Appleton & Com. New York.

(2) BAENA, J., et al, 1992. "Mapas Tectónico y Sismotectónico de España a escala 1:1.000.000". ITGE. Instituto Tecnológico y Geominero de España. Madrid. (In Spanish).

(3) ITGE, 1991. "Documentos sobre la geología del subsuelo de España". Instituto Geológico y Geominero de España. Madrid. (In Spanish).

(4) ITGE, 1992. "Proyecto de reprocesado e interpretación de varias campañas de geofísica aeroportada en el Macizo Hespérico ". Instituto Tecnológico y Geominero de España. Madrid. (In Spanish).

(5) CAPOTE, R., DE VICENTE, G. (1989). "Mapa geológico y tectónico" & "Mapa del Cuaternario en España". ITGE. Instituto Geológico y Geominero de España. Madrid (In Spanish).

(6) GARCÍN, M. et al. (1994). "Modellisation du Paléosite du Jarama". BRGM. Rapport N1092.

(7) GARCÍN, M. et al. (1994). "L'Europe de L'Ouest et la Peninsule Iberique de -120.000 ans a l'actuel. Reconstitutions cartographiques automatises". Colloque Geoprospective. UNESCO. Paris.

(8) IMBRIE, J. & IMBRIE, K.P. (1979) "Ice Ages: Solving the Mistery". McMillan. London.

(9) BARETTINO, D., BAJOS, C. (1994) "Paleoclimatiological Revision of Climate Evolution and Environment during the Quaternary on the Iberian Peninsula". Colloque Geoprospective. UNESCO. Paris.

FRENCH APPROACH IN PREDICTING THE EVOLUTION OF GEOLOGICAL ENVIRONMENT OF LONG LIVED RADIOACTIVE WASTE DISPOSAL

Aranyossy J.F., André-Jehan, B. Mouroux and M. Raynal

Agence Nationale pour la Gestion des Déchets Radioactifs (ANDRA)

Summary

The preliminary investigation phase currently executed in France under the responsibility of ANDRA, involves 4 districts located in three different geographical regions : in the North Eastern part, the "Meuse" and "Haute-Marne" districts correspond to the oriental limit of the vast sedimentary Parisian Basin ; in the South-Western part, the investigated zone of the "Vienne" district consists of a crystalline basement under a sedimentary cover ; in the South-Eastern region, the study area concerns the sedimentary formation of the Rhodanian Basin.

Taking into account this large scope of situations , the *"geoprospective"* approach should, in a first stage, consider the global role of the geodynamic processes responsible of the geological evolution of each study area. The basic principle of the *geoprospective* approach is to consider that the future evolution of a region is somehow dictated by a logical geological history. The method of investigation therefore consists in assessing and getting a comprehensive knowledge the past geologic evolution of the study areas to extrapolate it to the future at different time scale (thousands ; tens of thousand ; hundreds of thousand to the million of years...) relevant to the problem of radioactive waste geological disposal.

This knowledge will serve as the groundwork to the situations to be taken into account in accordance with the French Basic Safety Rule (RFSIII2f): (1) a *"reference situation"* corresponding to the future evolution of the repository and its environment as regards to events which are certain or highly probable within a period of 100 000 years after the repository closure ; (2) *"hypothetical situations"* corresponding to the occurrence of casual events, in addition to those of the reference situation, which may result in a preferential transfer of radionuclides between the repository and the biosphere.

Two major examples are given to illustrate this approach. The first concerns the future climatic evolution through the development of a new Würm type glaciation in northern Europe within the next 40 to 60 000 years. This evolution is considered as a part of the reference situation and should mainly be investigated in terms of its consequences on the groundwater circulation. The second exemple takes into consideration a foreseeable paroxysmic geodynamic event similar to the "Messinian crisis" which profoundly affected 5 millions years ago all the Mediterranean basin. This exeptionnal phenomenon is studied by ANDRA as an hypothetical situation to check its possible consequences on the erosion processes in the Rhodanien Basin.

Evaluation of Long-Term Geological Changes in Northern Switzerland and Effects on a Repository in the Crystalline Basement

M. Thury & W.H. Mueller

Nagra (Switzerland)

Abstract

In 1980, Nagra, the Swiss National Cooperative for the Storage of Radioactive Waste, started a comprehensive field investigation programme to assess the siting possibilities for a high level waste repository in the crystalline basement as well as the feasibility and safety of such a repository. In this context, the long-term geological changes and their effects on a potential repository were evaluated, supported significantly by the results of an extended neotectonic investigation programme. In the present abstract, the approach for this assessment, the various investigations and results, the qualitative and quantitative assumptions for long-term changes and their effects on a repository are briefly described. They are presented in detail in a geological synthesis report (THURY et al., 1994) which contains a reference list of 24 Nagra Technical Reports published on the subject of long-term changes.

The **approach** comprised the following steps:

- Analysis of the geologic evolution of Central Europe and, in more detail of Northern Switzerland.
- Assessment and quantification of geologic processes expected in Northern Switzerland.
- Performance of Neotectonic Investigations.
- Comparison of the results of neotectonic investigations with the conclusions of the analysis of the geological evolution.
- Derivation of processes and a set of conservative values for changes expected in the next million years.
- Evaluation of consequences on groundwater flow and on repository siting, design and long-term safety.

The analysis of the geological evolution resulted in two alternative scenarios, one with an ongoing alpine orogeny and one, in which the alpine orogeny is completed. Assuming, that the alpine orogeny goes on, the following **crustal movements** are expected in the next million years (in relation to a reference point in Basel):

– Crustal shortening in the Alps:	3 km
– Uplift of the Alps:	1,5 km
– Uplift of the Black Forest:	1 km
– Uplift of the selected siting area in the Tabular Jura:	200 m
– Horizontal dislocation of the molasse basin and Folded Jura (thrusted over the Tabular Jura):	500 m

The **Neotectonic investigation programme** comprised:

- Microearthquake monitoring
- Geodetic measurements
- Geomorphological studies
- Stress field measurements
- GPS measurements

For the assessment of the **microearthquake activity**, in 1983 a network of nine seismographs was installated. The low seismic activity in the selected siting area was confirmed: no earthquake with a magnitude above 1 (which is the detection limit of the seismic network) was recorded during ten years, while dozens of quakes of magnitude 1 to 4 were recorded in the neighboring Rhine Graben and Bodensee areas.

Geodetic measurements (precision leveling surveys) were repeated locally and analysed in the region. The mean vertical crustal movements of some groups out of 190 bench marks in different areas are in the order of 0,1 mm to 0,5 mm/y, which confirms the rates of vertical movements postulated above.

Geomorphological studies were carried out in order to detect active faults. The lineaments detected on satellite images were checked on aerial photos and in the field but no evidence of active faulting was detected. The levels and inclinations of different post-glacial terraces were analysed. Eventual transpositions of graved terraces along active faults are below the detection limit of a few meters. The evolution of the drainage pattern in the region since pleistocene was analysed. In the Black Forest young deviations of river courses confirm the ongoing uplift of this area. The levels of the deepest erosion channel of the Rhine river (actually filled with gravels) was analysed. Steps in this channel indicate block tilting along major faults of up to 10 - 20 m in 100'000 years.

Stress field measurements with the method of borehole wall breakouts in the Nagra deep boreholes confirm the actual stress field in the crystalline basement which was derived from earthquake analyses. It is similar to the tectonic stress field since Middle Miocene 15 - 10 Million years ago.

In 1988 a first double zero **GPS measurement** of 24 bench marks in different tectonic units in the region was carried out. In 1995, the measurement was repeated. The data are not yet analysed but it is expected that at least they confirm, that the postulated values for horizontal movements (decollement of the Folded Jura 500 m/million years or 0,5 mm/year) are conservative maximum values.

From the analysis of **climatic changes and erosion**, the following conclusions were derived for the next million years:

- Several glaciations are expected
- Climate can change to humid or arid
- Pronounced river erosion in regions with tectonic uplift (up to 200 m in the selected siting area) will occur
- Shift of river beds (the Rhine river, which is the discharge of the crystalline groundwaters from the siting area, may shift up to 2 km into the siting area) is expected
- Glacial erosion of quaternary sediments will take place
- No significant glacial erosion of bed-rock in the selected siting area is expected
- Denudation of 50 - 100 m can be expected

The **consequences** of all these expected long-term changes **on groundwater flow** in the crystalline basement of the selected siting area can be summarised as:

- Rise of groundwater levels in uplifted regions
- Sink and shift of groundwater exfiltration zone (Rhine river)
- No significant change of overall transmissivities of faults and of other water conducting features, but locally changes of interconnections of transmissive features
- Slight rise of regional hydraulic gradients but no significant change of flux in low permeable crystalline host rock between transmissive faults.
- In the part of the selected siting area that lies between the actual exfiltration zone (Rhine river) and the future exfiltration zone (shifted Rhine river), a change of groundwater flow field and possibly of hydrochemistry will occur due to the change of the exfiltration zone.

The **consequences** of all these expected long term changes **on repository siting and safety** can be summarised as follows:

- The maximum river erosion of 200 m does not affect the repository which will be placed deeper than 400 m
- The expected tectonic movements along major faults (faults with extensions of more than 1 km) of up to 100 m do not affect the disposal gallery panels if they are placed away from these faults
- Movements along small faults are expected to be less than 1 m and do not dangerously affect the bentonite barrier. The waste canisters should not be placed directly at the location of these small faults
- If the repository should be located in the part of the selected siting area that lies between the actual and the expected future exfiltration zone, further studies should reevaluate in detail the postulated movement of the exfiltration zone and the hydrodynamic and hydrochemical regime in this zone
- A change to arid climate would result in a smaller water flow in the Rhine river and a smaller dilution of potentially radionuclide containing groundwaters from the repository

The **conclusions** of the studies and investigations presented above are:

- Long-term geological changes in Northern Switzerland have been evaluated by studying the geological evolution and by neotectonic field investigations
- An understanding of the processes and coupled processes which can lead to long-term changes has been gained
- The processes have been quantified and conservative maximum values have been estimated for a time period of 1 million years
- The effects on a repository in the crystalline basement have been evaluated. Negative effects can be avoided by adequate repository siting and design
- Microearthquake monitoring will continue
- Further GPS-measurements are recommended
- Detailed studies in the selected siting area are recommended

Reference:

THURY, M., GAUTSCHI, A., MAZUREK, M., MÜLLER, W.H., NAEF, H., PEARSON, F.J., VOMVORIS, S. & WILSON, W. (1994): Geology and Hydrogeology of the Crystalline Basement of Northern Switzerland. Synthesis of Regional Investigations 1981 - 1993 within the Nagra Radioactive Waste Disposal Programme. Nagra Technical Report NTB 93-01, Nagra, Wettingen, Switzerland.

Characterization of Long-Term Geological Changes for Final Disposal of Spent Fuel in Finland

Paavo Vuorela
Geological Survey of Finland
Timo Äikäs
Teollisuuden Voima Oy, Finland
Runar Blomqvist
Geological Survey of Finland

Abstract

The bedrock of Finland is very old and major crustal deformation processes ceased long ago. At present continuous slow processes prevail and geological changes taking place today are very difficult to observe. Anticipated future geological changes are dominated by the renewed development of the continental ice sheet in northern Europe. The present climate will deteriorate to a state amenable to glacier formation. Continuous processes such as groundwater flow and interrelated hydrogeochemical phenomena will be influenced by changes in the climate as well as by developing permafrost. The crust itself will be loaded by the weight of the ice sheet, and will warp down.

The final disposal programme has been devised with even more exceptional future changes in mind. The process of site identification in the site selection research programme has been developed to consider the eventuality of future bedrock movements. Analysis of bedrock geometry and block patterns, together with related fracture zones assists in selecting a repository site where the risks of accumulation of large stresses, and their subsequent release as shear movements, can be minimized.

By studying the prevailing conditions and tracing the record of earlier events an understanding of the relevant processes in general is developed. Paleohydrogeology is one of the areas which can provide information relating to "why the conditions at the site today are as they are". Although it is not possible to predict the future behavior of a site in a detailed manner, it is possible to constrain the scenarios needed in the safety assessment by establishing and documenting real events that have sometimes occurred, and that will most probably be repeated.

1 Outline of Scenario on Geological Evolution

1.1 Structure and Geology

The Fennoscandian Shield together with the Ukrainian Shield are the exposed parts of the crystalline Fennosarmatian basement block which elsewhere in Europe is mostly covered by thick layers of Paleozoic or younger sedimentary rocks (Fig. 1). The Fennoscandian Shield can be geochronologically divided into different zones with decreasing ages from north-east to south-west. The Shield was formed and reworked in four main orogenies: the Saamian at about 3 Ga and Lopian at about 2.6 - 2.9 Ga in the northeastern part form the Archean basement, the Svecokarelian about 1.7-2.0 Ga in the central part and the Gothian 1.5 - 1.7 Ga in the southwestern part. The Caledonian 0.4-0.6 Ga orogeny caused basement reworking in the western and southwestern part of the Shield with some mild tectonic activity in the middle part of the shield, reflected by the presence of some carbonatite intrusions. Since that, no magmatic activity has been recorded.

The whole Fennosarmatian Shield is bordered on its southwestern side by a NW-SE striking deformation zone known as Tornquist-Teisseyre lineament (Franke et al. 1989). This zone was most active during the Paleozoic era. In the middle of the Fennoscandian Shield the boundary between the Archean and Proterozoic rocks is approximately parallel to this NW-SE direction, as is the related wide fracture zone in central Finland. Subparallel zones are also present in Finnish Lapland and the Kola Peninsula but their relative ages are very different. The main fracture zone in central Finland is most probably of Proterozoic origin, and apparently transects the entire crust (40 - 60 km). The main activating force in reactivating the younger lineament has been most probably the Mid-Atlantic ridge. The old deformation zones might also be well related to ancient plate-tectonic movements, as associated magmatic activity demonstrates that they are very old.

The Fennoscandian Shield is divided into large bedrock blocks bordered by wide ductile deformation zones, for instance (in Finland) the granite massifs of central Lapland and central Finland, which are bordered by deformed gneiss and schist belts. These larger blocks can then be subdivided into smaller domains, some hundreds of square kilometers in area, and still further into blocks tens of square kilometers in extent, approaching the scale appropriate for repository studies. To compile a prognosis of possible future bedrock movements in an area considered for permanent waste disposal, it is important to determine the regularity of significant bedrock structures that may be activated during future bedrock movements. Frequency distribution of the maximum orientations of the main fracture zones in Finland indicates both regularity and different ages of deformation (Tuominen et al. 1973, Talvitie 1977, Aarnisalo 1977, Vuorela 1982). Eight strike frequency maxima have been identified in both central and southern Finland, and schistosity, fractures and gravity anomalies all indicate statistically similar orientations.

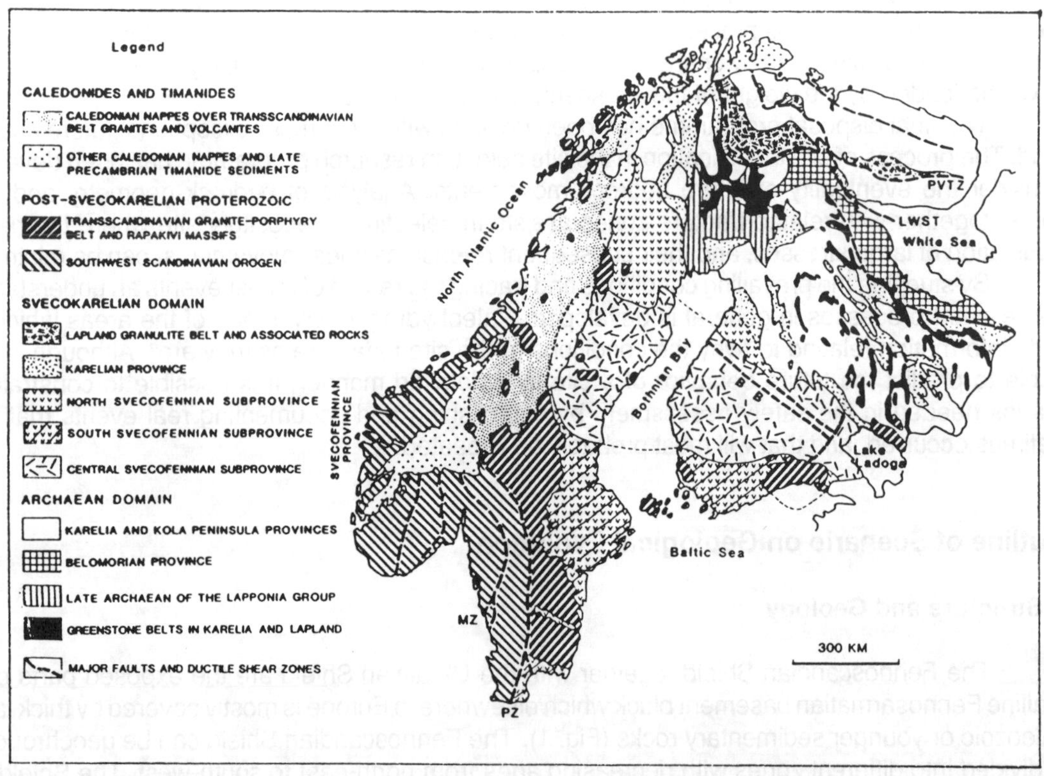

Figure 1. Scematic geological map of the Fennoscandian Shield (from Gorbatschev and Gaál 1987). LLB = Ladoga-Bothnian Bay Tectonic Zone: LWST = Lapland-White Sea Thrust Fault; MZ = "Mylonite" Zone; PBTZ = Pechenga - Varguza Tectonic Zone; PZ = "Prologine" Zone.

Bedrock blocks are represented in different size classes. Fracture zones can also be divided into classes of different size categories (Salmi et al. 1985). Strike frequency maxima of the main fracture zones often coincide with maxima of smaller fractures and even with schistosity. In a pattern of fracturing and block formation an important principle is to find significant size clusters which participate in bedrock movements as single, "unbroken" units. Granite and gabbro intrusions are often located at the intersections of the main fracture zones, such as the granite intrusions of the Veitsivaara, a previous investigation site, and the test drilling site Lavia in southwestern Finland. Intrusions have subsequently acted during bedrock movements as single blocks bordered by fracture zones.

1.2 Prevailing Continuous Processes

The Finnish bedrock is very old, and the current erosion level was achieved at an early stage, so that both uplift and erosion of the earth's crust has been slow during the last 1.7 Ga (Nurmi 1985). Apart from the eastern parts of Lapland, Finland has been above sea level for hundreds of million years (Donner 1976). The bedrock has of course responded in the course of time in accordance with the plate tectonics of the earth, and in this respect great age is not in itself an indication of a particular level of stability, but processes leading to major changes in the bedrock nevertheless ceased to affect this area some 300 Ma ago (Nurmi 1985). Finland today has no volcanic activity, no significant earthquake activity and no (anomalous) internal heat sources.

Bedrock movements today are very minor and can only be detected by sensitive instruments. Neither do current earthquakes present any particular seismic hazard. The only feature of the seismicity pattern clearly visible on both maps and in concentrations of epicenters is in the northeastern part of the country, in the Kuusamo region (Fig. 2). The largest magnitude earthquake recorded in Finland is 4.9. The evaluation of seismic risk based on the observations since 1880 has resulted in a magnitude extreme of 5.0 for the upper limit of magnitude (Saari 1992).

Land uplift is still in progress (Fig. 3) in the previously glaciated regions, implying that the land surface has not yet returned to a state of isostatic equilibrium. The current maximum annual rate of uplift on the western coast of the country is 8-9 mm. Observations suggest that the land uplift pattern is not completely regular (Kakkuri and Chen 1992). Land uplift somewhat in excess of 100 m can still be expected (Kakkuri 1986).

The horizontal crustal stress analyses suggest that the western part of the country seems to be under horizontal extension while the southeastern part is under compression. The maximum horizontal compression, in general, is oriented in a northwest–southeast direction (Chen and Kakkuri 1994).

The predominant geological processes operating in bedrock at present are those that proceed at even rate, i.e. gradual weathering and erosion at the surface, creep, or gradual adaptation to changes in stress state in the bedrock, infiltration of groundwater from the surface and groundwater flow.

The circulation of groundwater in bedrock is governed by precipitation, the topography of the land surface and the hydraulic conductivity of fractures. Of these properties the most rapid variations occur with respect to precipitation, the topography and fracturing being more permanent properties that only change over extended periods of time.

1.3 Expected Evolution

It is expected that the slow continuous processes operating today in the bedrock will show similar behaviour in the future even though the rate of processes affecting bedrock movements or groundwater flow might change. Variations in the rate of these processes, or in the distribution of the bedrock properties influencing these processes, will be mostly affected by external climatic evolution. Evidence from previous glaciations lends justification to the assumption that the climate and geosphere will continue to interact in a similar manner in the future. With respect to the climatic history recorded in the sea-floor sediments, for example, the present warm period is more or less a transient disturbance in a more dominant cold and harsh environment.

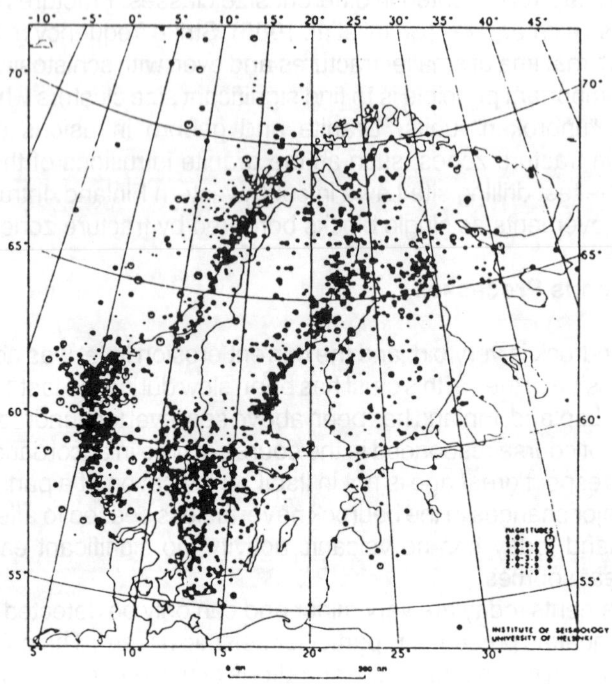

Figure 2. Earthquake epicenters in northern Europe in 1965 - 1989 (from Korhonen and Ahjos 1984).

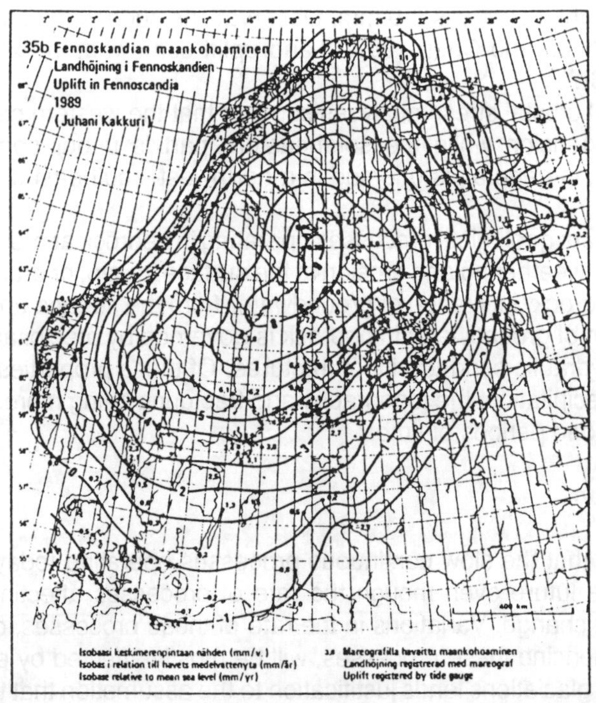

Figure 3. The observed land uplift in the Fennoscandian Shield in mm/year. Isobases are relative to the mean sea level (Kakkuri 1991).

Based on the scenario compiled by Ahlbom et al. 1991, the climate in Scandinavia will gradually become colder (Fig. 4). With time the changes in the climate will permit the growth of an ice sheet in about 5000 years. The colder climate also means the development of permafrost, and a drier climate, in general. Due to the weight of the ice sheet the crust will warp downwards but sea-level will still fall. After a minor and somewhat warmer period the stadial conditions will develop again at around 20 000 years from now. The ice sheet will grow thicker and reach a thickness of more than 1 km. Before full glaciation a further interstadial will prevail at around 30 000 – 50 000 years. However, at that time periglacial conditions similar to those in Greenland or Antarctic today are anticipated. After this, the ice sheet reaches its expected maximum at about 60 000 years, when it will be 3 km thick and cause a crustal downwarping of 0.7 km. The gradual melting of the ice sheet with rapidly alternating interstadial/stadial conditions will result in a similar condition to our interglacial after 120 000 years (Ahlbom et al. 1991).

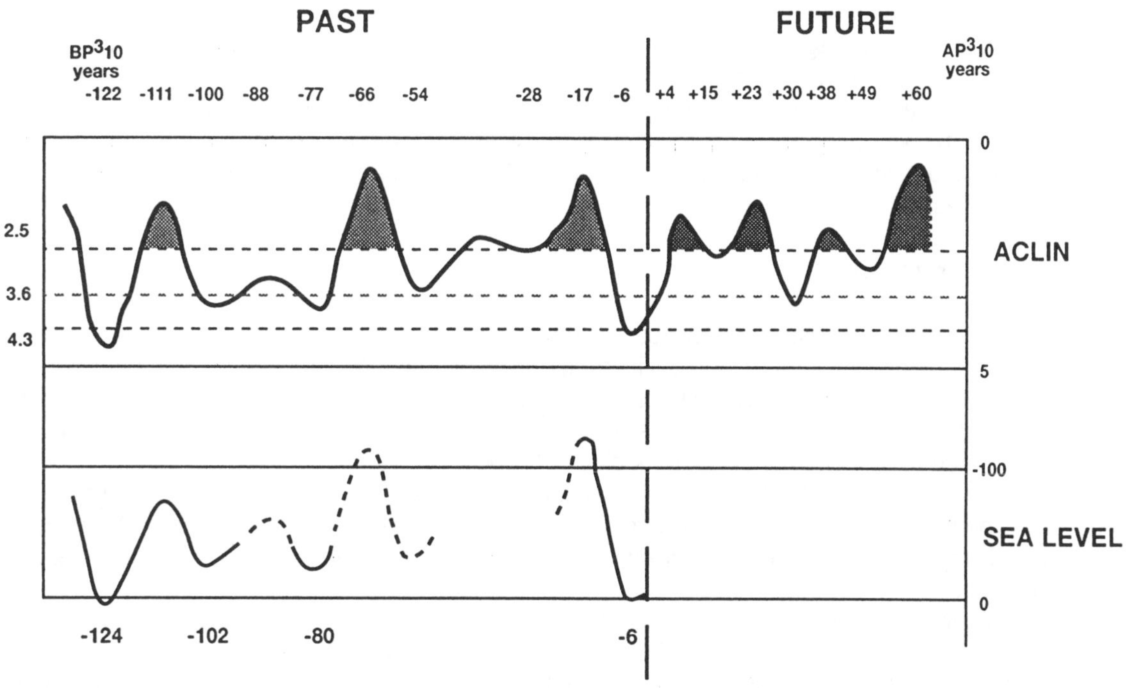

Figure 4. The last glacial cycle and the ACLIN projected climate of the next 60 000 years (from AECL 7822).

2 Approaches to Evaluate and Predict Future Changes

2.1 Long-term Stability of the Bedrock

Most of the instability and activity in the earth's crust occur in active spreading zones at ocean ridges and at destructive continental plate margins. In the shield areas such processes are absent or very slow. The active zones do however affect the stress field in the Canadian and Fennoscandian Shields where considerable horizontal compression is encountered, as well as phenomena due to glaciations. The regular orientation distribution of crustal deformation zones established worldwide, Finland included, is caused by the large scale regional stress field (Stephansson et al.1987). An important additional factor in the Fennoscandian Shield is land uplift, which is surely one reason for the prevailing regional stress, as the subsided earth's crust is restored to its former shape.

Since the last glaciation, uplift in Finland has been about 700 m at the maximum with movements between separate blocks accounting for no more than 10 %. The highest known postglacial fault-scarp amounts to 35 meters and is located in northern Sweden. Seismic activity in the Fennoscandian Shield has been related to melting stage of the previous ice sheet. The next critical time period will be immediately after the next ice age, in approximately 100 000 years. Present day bedrock movements are at a very low level. Younger tectonic movements, faults and fracturing, are controlled by the existing zones of weakness in the bedrock. This is also the main reason for the low level of seismic activity in Finland, in that new failure of the bedrock is not necessary, since stresses are effectively released along old fracture zones. This also means that the old fracture zones are often in more or less constant very slow motion, as indeed is indicated by the abundance of weak seismic events (Slunga 1989). The block-mosaic structure of the bedrock and block movements associated with it are very important factors in predicting the long term behavior of the crust and the intergrity of proposed final repositories. By monitoring the current movements it is possible to forecast the location and direction of future movements, should their magnitude and rate increase as could be anticipated following an ice age.

The behaviour of the block structure model of the bedrock can be evaluated, to a certain extent, by analysing the spatial distribution of the younger faulting and by following the present-day bedrock movements. That is why evidence of postglacial faulting and present-day bedrock movements have been under intensive study in co-operation with a number of different research institutes in Finland and other northern countries.

Leveling measurements

The idea of a domelike smooth isostatic uplift of the shield has been also confirmed by precise leveling studies which have been carried out in three phases in Finland for more than one hundred years, the first being between 1892 - 1910, the second between 1935 -1955, concluding in northern Finland in 1955-1975, and the third starting in 1978 in central Finland, and which is now nearing completion. In detail the uplift is related to movements of bedrock blocks (Veriö et al. 1993).

To analyse slow bedrock movements, which usually take place by creep, a method of repeated leveling has been used. The density and accuracy of the national leveling network is so high that the relevelings of properly selected leveling lines will give us a comprehensive picture of bedrock movements in the 20th century.

Measurement lines have been established across known lineaments interpreted at a scale 1:1 000 000 (Vuorela 1974), which usually indicates significant fracture zones in the bedrock. Releveling measurements were made on 42 lines by the National Board of Survey. The intervals of relevelings varied from 9 to 66 years. Six lines have been measured three or more times, which has facilitated determining whether or not the direction and rate of the observed movement has been continuous (see Fig. 5 and Fig. 6).

The results of 42 releveled lines indicated a significant change in local height, on 20 lines, whereas the other 22 lines indicated immobility or slow tilting in accordance with the assumed present-day bedrock uplift.

Intersecting fracture zones were interpreted in detail from topographic maps in the scale 1: 50 000. Fractures were classified into four categories according to size and the releveling results were compared against the fracture interpretations. The precise locations of the faults were nevertheless impossible to determine because a lineament between benchmarks actually consists of a complicated zone of bedrock fractures.

Horizontal movements

Vertical movements have been monitored over long time periods in Finland. Horizontal movements, however, are not so well known. A present assumption is that horizontal movements in the bedrock are quite comparable in size with the vertical ones. Horizontal strike slip fault movements can also be measured by present methods. If a fault zone is steep and thus related to a straight lineament,

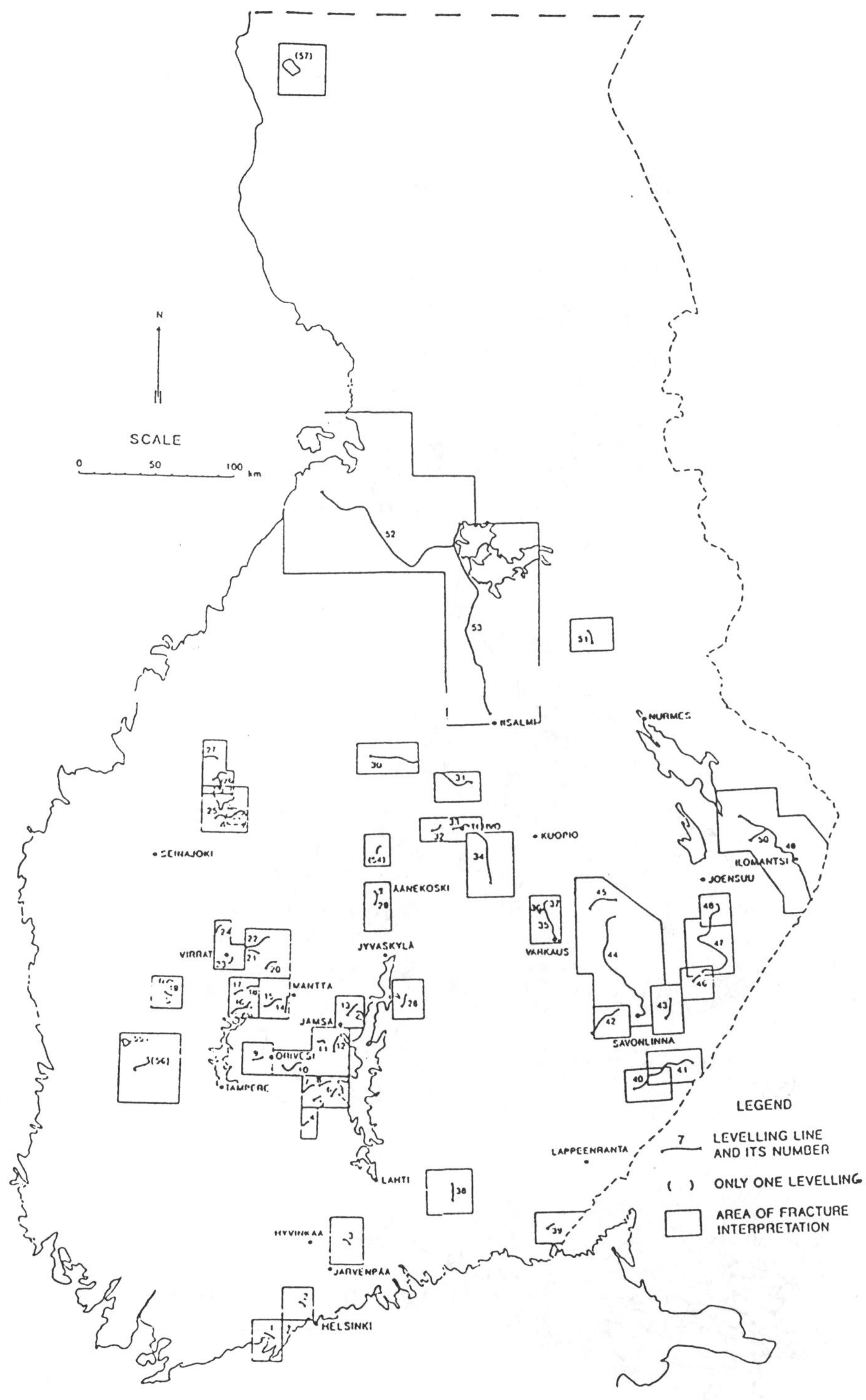

Figure 5. The areas of the levelling programme of fracture zones in Finland (from Veriö et al. 1993).

Figure 6. The areas with significant altitude change in the fracture levelling programme (from Veriö et al. 1993).

benchmarks can be established and measurement done by GPS method (Global Satellite Positioning), for instance. Low angle or subhorizontal faults are more complicated. Since they are often poorly indicated on the bedrock surface and they are generally difficult to locate. By using regional measurements crustal deformations can be identified as Kakkuri and Chen (1992) have shown in Finland using the observations of the national first order triangulation (1991). On the basis of two measurement campaigns in central Finland 1991 Chen has compiled a strain map of Finland (Fig. 7), showing that certain areas are under compression and some are in extension. This preliminary result offers a basis for further more detailed measurements concerning movements and the state of stress in the bedrock.

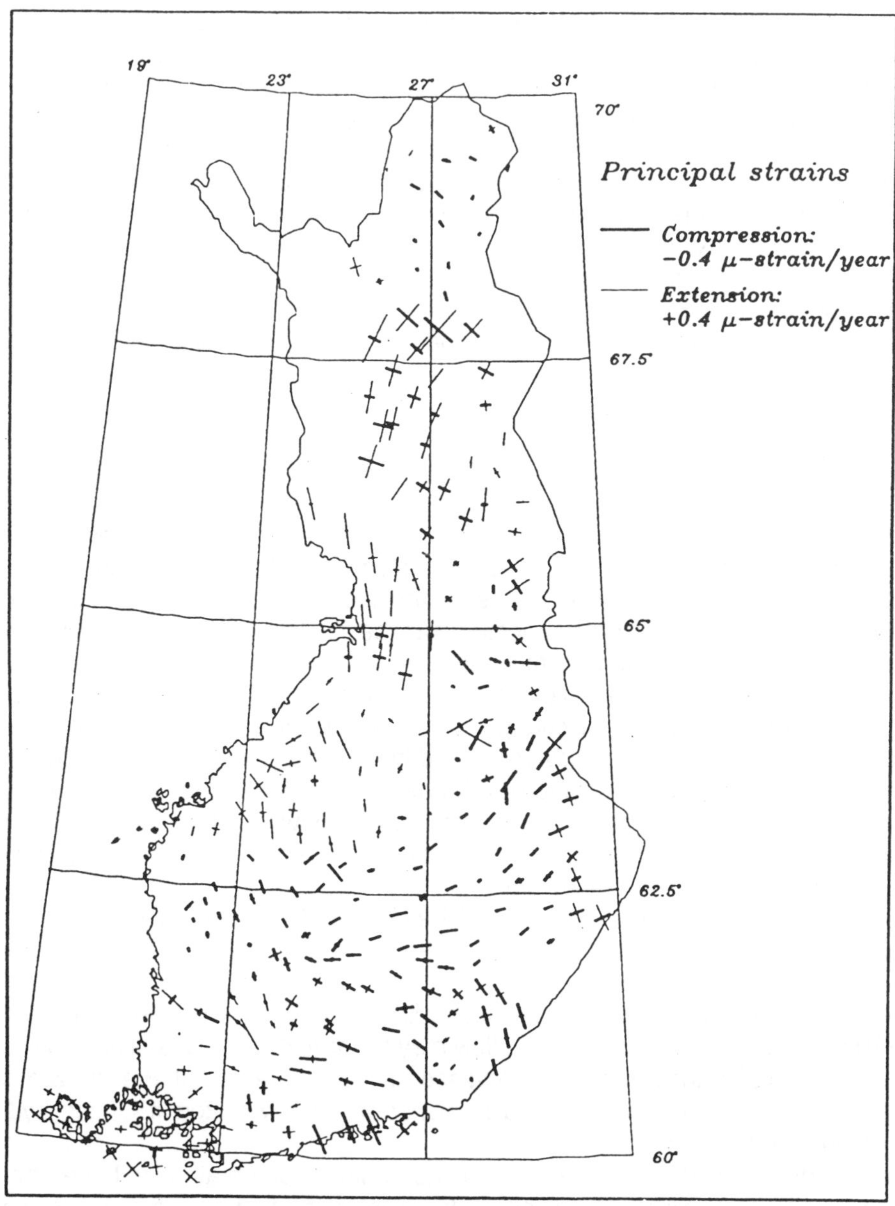

Figure 7. Strain patterns in Finland derived from the observations of the first-order triangulation (Chen 1991).

The GPS method will also be used to monitor bedrock movements in Finland. Five GPS stations have been established by the Finnish Geodetic Institute to cover the whole country, and seven more stations are planned. Local grids are being constructed at the three investigation sites for the disposal of the spent nuclear fuel.

Postglacial faults

Seismic results, leveling and geodetic measurements seem to support the idea of differential block movements. The present slow movements within the bedrock are concentrated in fracture zones, as it is believed to be the case with faults that will be activated during the accelerated rate of movements just after the melting of a future ice cap. In the northern part of the Fennoscandian Shield several steep fault scarps have been identified (Tanner 1930, Kujansuu 1964, Lagerbäck 1983, Olesen 1985). The possibility of having a new fault develop in previously intact bedrock has resulted intensive study. The relationship between fault scarps and older fracture zones in the bedrock is not quite clear. Fault scarps often show, as with the Pasmajärvi fault in Finland, repeated brecciation. Movement seems to be indeed postglacial but this does not explain the origin of the whole fault zone. The NE-SW orientation of the fault scarps is perpendicular to the prevailing old faulting trend. Leveling network and GPS benchmarks have been assembled at Pasmajärvi to detect movements in the postglacial fault scarp and in the older fault zone. An interesting possibility is that the steep fault scarps may turn into subhorizontal zones at depth in the bedrock (Talbot et al. 1989).

In Russian Karelia steep fault scarps have been reported as indications of possible postglacial movements. The strike of the structures coincide with the orientation of fracture zones in Finland, and these phenomena will be carefully studied.

2.2 Changes in Groundwater Flow

Before the next glaciation no significant changes in the hydraulic properties of the bedrock are expected. Accommodation to stress pattern and dissolution/precipitation of fracture filling material will sporadically alter the properties of single fractures but the overall properties will remain more or less as today. The greatest changes will occur in connection with the glaciation. The main effects will most likely be changes in gradient, rather than in transmissivity.

In the early stages of a glaciation cycle temperature and precipitation will decrease and permafrost will most likely develop. These changes will, in general, to a great extent reduce groundwater recharge. This again will mean that groundwater flow driven by the topographical gradient will diminish or even cease entirely. During the deglaciation discharge is much greater than at present. In deglaciated regions the ground surface is the highest level to which the groundwater table can rise. This would mean that the maximum hydraulic gradient would follow the topography of bedrock, assuming that the importance of glacial soil formations is negligible.

It is somewhat uncertain how great the infiltration rate during the glaciation would be beneath the ice sheet itself. Typical melting rates vary between 2-50 mm/y at base of a large glacier. The infiltration rate, as well as, the piezometric head can vary considerably in different parts of an ice sheet (Ahlbom et al. 1991). The groundwater flow during the glaciation will be driven by the gradient caused by thickness and form of the ice sheet. Local variations in the flow may be large due to transmissivity variations in the underlying bedrock as well as the existence of crevasses in the ice sheet itself. Pressurized water bodies below the ice sheet may also cause some local disturbances in the flow pattern if surging, hydrofracturing or other types of instability occur.

The changes caused by the ice front, or the variable temperature at base of a glacier are "temporary" in their nature. The longest time period for which retreating ice remained its position during the last ice age was around 1000 years, and this left a remarkable end-moraine, known as Salpausselkä behind.

The chemical properties of the present groundwater, such as high salinity and C-14 values suggest rather restricted flow and very slow turnover of groundwater. These observations may, however, be biased towards the situation during the latest glaciation(s), and conclusions concerning today's flow pattern may not be derived directly from these results. Therefore to understand how salinity, infiltration, transmissivity (distribution, structures) and land uplift interact study program has been commenced. By simulating time-dependent land uplift by a three-dimensional flow model of an investigation site the salinity and pressure field can be studied (Löfman and Taivassalo 1992).

2.3 Paleohydrogeology

Paleohydrogeological methods can be helpful in attempting to forecast eventual future changes in the hydrogeological regime of a potential disposal site. For the time being paleohydrogeology, mostly because of the relatively sophisticated techniques and extensive sampling programmes needed for obtaining qualified results, has received relatively modest attention within site characterization programmes.

In acquiring such information all the available sources ought to be used. So far both fracture-minerals and groundwaters have been utilized. Analytical tools include various microanalytical and isotope methods (e.g. ^{13}C, ^{18}O, $^{87}Sr/^{86}Sr$, $^{234}U/^{238}U$), dating (U/Th method) and residence time measurements (^{3}H, ^{14}C). In Finland examples from a few sites have already been described.

At Outokumpu in SE Finland a major fracture zone was sampled for groundwater (Blomqvist et al. 1989; Ivanovich et al. 1992). The fracture water at a depth of 400 m proved to possess an exceptionally light ^{18}O value. Compared to isotopic values of fresh groundwaters in the vicinity, the delta ^{18}O values of the sampled fracture-water was about 3 % lighter, indicating a distinctly colder conditions during recharge. The fracture-controlled water is interpreted as a mixture of glacial melt-water with the local dominant chloride water (Blomqvist et al. 1994). The implication is that the recharge of glacial fresh water reached a depth of about 400 m.

At Olkiluoto in Eurajoki, one of the potential sites studied for the final disposal of high-level waste, euhedral calcite was sampled from drill-cores representing depths from 613 and 760 m (Blomqvist et al. 1992). When dated with the U/Th method, the crystal-shaped yielded young (< 300 ka) ages. The youngest calcite type which was located next to the fracture, yielded an age of 120 ka, which coincides with the previous interglacial period. A variety stable isotope information was obtained from the young calcite types. The correlation of the data with isotope data from the ambient groundwater, provides information on the changes of the hydrogeological regimes since the formation of the particular calcite.

Unfortunately the dataset was partly incomplete and the final analysis of the results are yet to be done. However, the first implication of the isotope results is that the hydrogeochemical conditions 120 ka ago and the present ones are relatively similar.

At the national analog study site at Palmottu, Nummi-Pusula, strong evidence exists for a glacial melt-water recharge in the bedrock. Interestingly, the groundwater body which yields the cold ^{18}O signature is a sulfate water (750 mg/l) overlain by dilute fast moving fresh groundwaters. This implies that the sulfate in the water was oxidized since the intrusion of the cold-climate water. It is likely that oxygen in the meltwater was involved in the process, but based on calculations, additional oxygen would also have been needed.

3 Consideration of Changes in Safety Analysis

It is difficult to confident that geological processes will take place in a certain predictable and predetermined manner, even though the evidence from recent geological history suggests similar processes will also occur in the future. On the basis of published studies, (TVO, 1992a) it seems that in Finland, and within the Fennoscandian Shield in general, bedrock properties, such as the distribution of hydraulic conductivity, are more or less uniform. Hydrogeochemical and isotope data suggest, however, that areas in different parts of the country have undergone different processes or show different manifestations of the same processes. As long as it is difficult to evaluate the importance of possible changes it has been considered advisable to develop scenarios and study the uncertainty caused by possible changes through "what if" -type cases in the safety analysis (Vieno et al. 1992, Teollisuuden Voima Oy 1992b).

The minor changes of the hydraulic conductivity have typically been considered in the safety assessment by selecting generally conservative parameters. The safety assessment, when analysing transport, is not restricted to any particular transport route but also studies pessimistic cases for radionuclide transport. These changes are irrelevant as long as the canister stays intact and other barriers perform as assumed, so that before geological changes become important regarding safety targets, they must be capable weakening the barriers, and finally breaking the canister either mechanically or chemically, by accelerated corrosion for example. Cases with oxidizing conditions, as well as with variable salinities have been studied with respect to solubilities (Vieno et al. 1992). The flow effects of the saline groundwater have also been considered. The disposal plan can be regarded as acceptable in all "what if"-type scenarios reviewed.

Post-glacial faulting has been considered as one mechanism potentially capable of breaking one or more canisters. There is, however, as yet no scientific basis for estimating its probability. As an extreme case, without considering the probability of this kind of movement the effects of the possible fault has been analyzed in safety analyses (Vieno et al. 1992). For this scenario it was assumed that a canister will fail due to faulting at either 10 000 or 1 000 years which was regarded as conservative (see Chapter 1.3). To increase the conservativeness oxidizing conditions were assumed to prevail in the bentonite and bedrock. The occurrence of a major rock shear taking place precisely at the selected site and intersecting the deposition hole is very unlikely, and an event that could cause the simultaneous failure of all 60 canisters in a single disposal tunnel, for example, seems even less probable. The analysis showed that the safety of the final disposal site was not seriously impaired (Vieno et al. 1992, Ruokola 1994).

4 Concluding Remarks

The main task of studying long-term changes must be recognized as the need to understand the long-term behavior of the bedrock in general. The overall purpose of this is to develop models for the future behavior of the geosphere, but how reliable these models will be is concern for the safety assessment. The attempts to characterize long-term changes in bedrock should also allow the introduction of estimations of uncertainties into sensitivity analysis, for example, in the safety assessment.

Within a large shield area like the Fennoscandian Shield the consideration of long-term changes at a specific site is difficult. Since it is necessary to deduce whether the conditions met today are the result of the latest glaciation, a mixture of several earlier glaciations or merely the result of processes initiated after the retreat of the last ice sheet. The processes and their results in the bedrock are somewhat tractable but will the processes at the same place repeat themselves as before?

What are the consequences if future changes cannot be characterized by a site-specific approach? To discard any of the processes that might potentially cause significant, safety-related changes is difficult. Most probably the technical optimization of the disposal concept becomes difficult or impossible. Uncertainties regarding processes, though questionable and improbable, have to be incorporated within the performance assessments of technical barriers.

Some of the uncertainties, such as bedrock movements for example, can be evaluated to some extent in the site selection research programme by considering the method for initial site identification. Site identification in Finland is based on the model assumption of a block pattern in the bedrock (Fig. 8). In this mosaic-like pattern bedrock blocks of different degree can be identified by interpretation of satellite images, geophysical and geological information and topographical maps (Salmi et al. 1985). The safety target is to locate less fractured blocks bounded by significant zones of weakness. This model assumption can be developed to accommodate different principles of structural analysis, such as fractal analysis. The block structure developed in the bedrock early on, at least hundred millions years ago. Some evidence suggests that the stresses attempting to cause bedrock movements are released by creep processes in large fracture zones. Since the size of the blocks and degree of the fracture zones vary the most preferential in the site selection are those blocks comprising rather intact bedrock and located inside a larger bedrock block.

Since the long-term safety of permanent disposal cannot be demonstrated by testing, the only method is to rely on calculations based on solid assumptions. These assumptions should, as far as possible, be tested by experiments or analogies in nature. Some processes, however, may be more important to planning of the disposal concept than assessing long-term safety. At least the possible maximum intensities such as changes in the stress state (location and orientation of tunnels) and groundwater chemistry (selection of materials) should be evaluated.

It is evident that future evolution does not jeopardize site suitability and the safety of final disposal if implemented according to the plan presented in 1992 (Teollisuuden Voima Oy 1992b). A better understanding of the processes influencing the disposal environment can provide possibilities for technical optimization of the disposal concept. More important, however, is to understand these processes in order to build confidence in the ability to predict future evolution and thus accommodate an adequate safety margin in the disposal concept at the selected site.

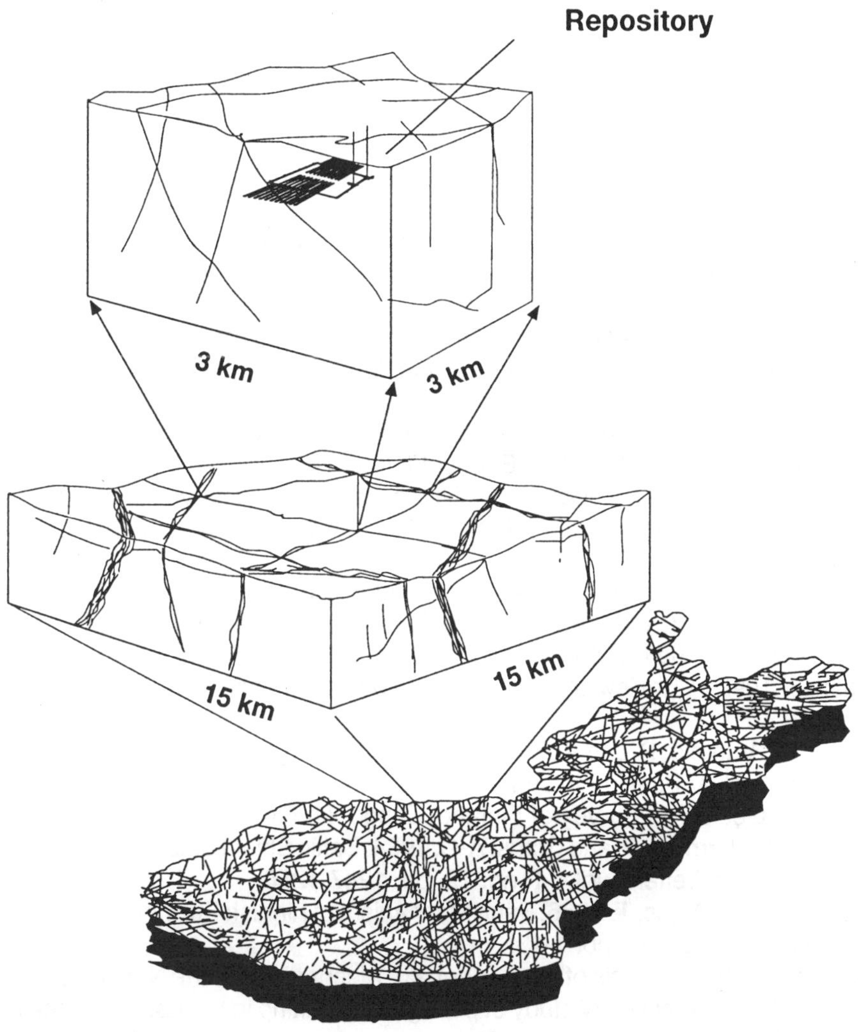

Figure 8. Geological approach selected for locating possible areas for site characterization (Teollisuuden Voima Oy 1992a).

REFERENCES

Ahlbom, K., Äikäs, T., and Ericsson, L. 1991. SKB/TVO Ice Age Scenario. Helsinki. Nuclear Waste Commission of Finnish Power Companies, Report YJT-91-19.

Chen, R. 1991. On the Horizontal Crustal Deformation in Finland. Reports of the Finnish Geodetic Institute, 91:1, 98 p.

Chen, R. and Kakkuri, J. 1994. Feasibility Study and Technical Proposal for Long-term Observations of Bedrock Stability with GPS. Finnish Geodetic Institute. Nuclear Waste Commission of Finnish Power Companies, Report YJT-94-02.

Franke, D., Kölbel, B. and Schwab, G. 1989. Zur Interpretation der Tornquist-Teisseyre Zone nach plattentektonischen Aspekten. Z. Angew. Geol., 35, 193-198.

Gaál, G. and Gorbatschev, R. 1987. An Outline of the Evolution of the Baltic Shield. Precambrian Res. 35, pp. 15-52.

Kakkuri, J. 1986. Newest Results Obtained in Studying the Fennoscandian Land Uplift Phenomenon. Tectonophysics, vol. 130.

Kakkuri, J. 1991. Geodetic Operations in Finland 1987-1991. Helsinki.

Kakkuri, J. and Chen, R. 1992. On Horizontal Crustal Strain in Finland. Bulletin Geodesique, 66, pp. 12-20.

Korhonen, K. and Ahjos, T. 1984. A Catalogue of Historical Earthquakes in Fennoscandian Area. Institute of Seismology, University of Helsinki.

Kujansuu, R. 1964. Recent Faults in Lapland. Geologi 16, No 3. (in Finnish)

Lagerbäck, R. 1988. Postglacial Faulting and Paleoseismicity in the Lansjärv Area, Northern Sweden. SKB Report 88-25.

Löfman, J. and Taivassalo, V. 1992. The Influence of Salinity on Groundwater Flow. Technical Research Centre of Finland, Nuclear Engineering Laboratory. TVO/Site Investigations, Work report 92-88. (in Finnish)

Nurmi, P. 1985. Future Long-term Environmental Changes in Finland and Their Influence on Deep-seated Groundwater. Geological Survey of Finland, Nuclear Waste Disposal Research. Report YJT-85-21. (in Finnish)

Olesen, O. 1985. Postglacial Faulting of Masi, Finnmark, Northern Norway. Abstract: XV Nordiska geofysikermötet 15.-17.1.1985, Espoo, Finland.

Paananen, M. 1987. Geophysical Studies of the Venejärvi, Ruostejärvi, Suasselkä and Pasmajärvi Postglacial Faults in Northern Finland. Nuclear Waste Disposal Research, Report YST-59. (in Finnish, abstract in English)

Proceedings from Workshop on Transitional Processes, 1984. AECL-7822. Canada.

Ruokola, E. (editor) 1994. Review of TVO's Spent Fuel Disposal Plans of 1992. Finnish Centre for Radiation and Nuclear Safety. STUK-B-YTO 121, July 1994, Helsinki.

Saari, J. 1992. A Review of the Seismotectonics of Finland. Imatran Voima Oy. Report YJT-92-29.

Salmi, M., Vuorela, P. and Kuivamäki, A. 1985. Geological Site Selection Studies for the Final Disposal of Spent Nuclear Fuel in Finland. Nuclear Waste Commission of Finnish Power Companies, Report YJT-85-27, Geological Survey of Finland, Espoo. (in Finnish)

Slunga, R., Norrman, P. and Glans, A-C. 1984. Baltic Shield Seismicity, the Result of a Regional Network. Geophys. Res. Letters, vol. 11, no. 12, pp. 1247-1250.

Slunga, R. and Ahjos, T. 1986. Fault Mechanism of Finnish Earthquakes, Crustal Stresses and Faults. Geophysica, vol. 22, 1-2, pp. 1-13.

Slunga, R. 1989. Analysis of the Earthquake Mechanisms in the Norrbotten Area. In: Bäckblom and Stanfors (eds.); Interdisciplinary study of post-glacial faulting in the Lansjärv area northern Sweden 1986-1988. SKB Technical Report, TR 89-31.

Talbot, C., Munier, R. and Riad, L. 1989. Reactivations of Proterozoic Shear Zones. In: Bäckblom and Stanfors (eds.); Interdisciplinary study of post-glacial faulting in the Lansjärv area northern Sweden 1986-1988. SKB Technical Report, TR 89-31.

Tanner, V. 1930. Study of Quaternary System in the Northern Fennoscandia IV. Bulletin de la Commission Geologique de Finlande 88, 594 p. (in Swedish)

Teollisuuden Voima Oy 1992a. Final Disposal of Spent Nuclear Fuel in Finnish Bedrock; Preliminary Site Investigations, Teollisuuden Voima Oy, Helsinki 1992. Nuclear Waste Commission of Finnish Power Companies, Report YJT-92-32E.

Teollisuuden Voima Oy 1992b. Final Disposal of Spent Nuclear Fuel in Finnish Bedrock; Technical Plans and Safety Assessment, Teollisuuden Voima Oy, Helsinki 1992. Nuclear Waste Commission of Finnish Power Companies, Report YJT-92-31E.

Tuominen, H.V., Aarnisalo, J. and Söderholm, B. 1973. Tectonic Patterns in the Central Baltic Shield. Bull. Geol. Soc. Finland 45, pp. 205-217.

Veriö, A., Kuivamäki, A. and Vuorela, P. 1993. Recent Displacements within Reactivated Systems in Finland; Results and Analysis of Levelling Measurements Carried out by the National Land Survey of Finland between 1974-1992. Geological Survey of Finland, Nuclear Waste Disposal Research, Report YST-84. (in Finnish, abstract in English)

Vieno, T., Hautojärvi, A., Koskinen, L. and Nordman, H. 1992. TVO-92 Safety Analysis of Spent Fuel Disposal, Helsinki 1992. Nuclear Waste Commission of Finnish Power Companies, Report YJT-92-33E.

Vuorela, P. 1982. Crustal Fractures Indicated by Lineament Density in Finland. Photogr. J. of Finland, vol. 9, no. 1.

Vuorela, P., Kuivamäki, A. and Veriö, A. 1991. Observations on Long-term Stability of the Bedrock in Finland. Proceedings of OECD/NEA Workshop on "Long-term Observation of Geological Environment: Needs and Techniques", 9th - 10th September 1991, Helsinki, Finland.

Analysis of Long-Term Geological and Hydrogeological Changes in the Swedish Programme for Final Disposal of Nuclear Waste

Lars O. Ericsson
Swedish Nuclear Fuel and Waste Management Co, SKB
Stockholm, Sweden

Geoffrey S. Boulton
Dept. of Geology and Geophysics, The University of Edinburgh,
Edinburgh, U.K.

Thomas Wallroth
Dept. of Geology, Chalmers University of Technology,
Göteborg, Sweden

Abstract

In assessing the safety of deep disposal of nuclear waste in crystalline rocks it is important to establish whether recent or future changes in loading can lead to fracturing and block displacement which may change the hydrogeological setting of a repository. Furthermore, it is of vital importance to understand how future climate changes, especially future glaciations, will influence the groundwater flow around a deep repository.

The Swedish programme comprises R&D activities which attempt to quantify probable impacts of earthquakes, glaciation and land uplift. These activities emphasise geodynamic processes in the Baltic Shield, postglacial faulting and glacial impacts on hydrogeology and ground water chemistry.

A time-dependent, thermo-mechanically coupled, three-dimensional model of the ice sheet behaviour in Scandinavia has been developed. The model is driven by changes in the elevation of the permanent snowline on its surface and by air temperature and predicts the behaviour of the ice sheet for an earth's surface of given form and mechanical properties.

The ice sheet model reconstructs the ice sheet thickness, ice sheet temperature distribution, including basal temperature, basal melting pattern and velocity distribution. The model is coupled to a subglacial Darcian groundwater flow model which in turn provides boundary conditions for evaluations of long-term hydrogeological evolution at specific sites.

INTRODUCTION

Geologically speaking, Sweden is situated within the Baltic Shield, which consists mainly of ancient rocks more than 900 million years old. The dominant crystalline bedrock consists of granites and gneisses. It is anticipated that spent nuclear fuel in Sweden will be deeply buried in these rocks. A deep repository will depend upon engineered barriers and a natural geological barrier to isolate the waste for a long enough period of time to permit radioactive materials to decay to natural background radiation levels. Direct radiation from the waste can be prevented by an intact crystalline rock block several metres thick. The only way in which waste deposited deep in the rock mass can affect humans, animals and plants is therefore through transport of radionuclides from the repository with the groundwater (or gas) in the rock to the biosphere. This transport can only take place through the fracture systems in bedrock. In general, the groundwater flux in fractured crystalline bedrock in Sweden is very low at such levels, and for this reason the repository is planned for about 500 m below the ground surface.

Bedrock has a number of fundamental properties that are being exploited for the long-term performance and safety of the repository. These are:

- Mechanical protection
- Chemically stable environment
- Small groundwater flux

These properties are more or less coupled to each other through physical or chemical processes.

EXPECTED GEOLOGICAL EVOLUTION OVER THE NEXT 100 000 YEARS

It is important to establish whether recent or future bedrock movements can lead to new fracturing, and whether load changes or rock block displacements can decisively alter the hydrogeological situation around a final repository. General objectives of the programme are therefore to:

- quantify or set limits on the consequences of earthquakes, glaciation and land uplift in analysing the safety of a final repository for spent nuclear fuel, and

- process, evaluate and increase knowledge concerning geodynamic processes in the Baltic Shield.

Tectonics and Seismic Activity

Tectonics is a collective term for study of the deformation of the earth's crust and the structures, from the millimetre to the kilometre scale, which result from it. It is essential to understand the brittle tectonic evolution of the Baltic Shield. The stress situations that have arisen during continental drift, "the palaeostress field", can be roughly analysed from the occurrence, orientation and age of dyke systems and the minerals precipitated in fractures. It is further possible to use isotopic data to deduce changes in former vertical loads such as might occur when overlying sedimentary strata are eroded away. Successive glaciations have also changed the load on Swedish crystalline basement. As a consequence of these varying orientations and magnitudes of stress, Swedish bedrock has acquired geometrically complex patterns of fractures. It is unlikely that new fractures will be developed in the current tectonic regime, although reactivation of existing fracture zones or faults may occur.

It is also unlikely that the present regime of passive response to ocean floor spreading in the Mid-Atlantic will change substantially within the near future or within the next 100 000 years (Muir-Wood, 1993; Larsson & Tullborg, 1993). We must however consider the tectonic impact of future ice loading, which will influence the stress in the upper crust within this time period (Ahlbom et al., 1991; Björck & Svensson, 1992; Eronen & Olander, 1990).

Seismic activity in the Baltic Shield is mainly controlled by the plate-tectonic processes and ongoing land uplift. The results of seismic measurements (Slunga, 1985; Slunga & Nordgren, 1987; Slunga, 1989) show that most of the stress that causes earthquakes has a compression direction of N60W, which is approximately perpendicular to the plate movement driven from the Mid-Atlantic Ridge. The global database of the "World Stress Map Project" also shows relatively good agreement between the plate tectonic movement of the earth's crust and the greatest horizontal principal stresses (compression). Certain deviations occur in the stress field in the Baltic Shield, which can be explained as a consequence of the glacio-isostatic uplift (Muir-Wood, 1993).

If deglaciation is relatively rapid, the earth's crust is subjected to regional stress differences which can trigger movements along pre-existing zones of weakness. The deglaciation phase following the most recent glaciation was significantly faster in northern Sweden than in the southern parts of the country. This is considered to be an important cause of the neotectonic and postglacial movements that have been inferred in, for example, the Lansjärv area in Lappland of northern Sweden (Muir-Wood, 1993).

Glaciation scenario

During 1990-1991, SKB and Teollisuuden Voima OY (TVO) in Finland carried out a joint inventory of the international state of knowledge regarding ice ages. The purpose was to describe when future ice ages can be expected and what changes in the geosphere occur in connection with them (Ahlbom et al., 1991).

During the past 20 years a large quantity of detailed data have been collected which support the Milankovitch (1941) theory, in which global climatic change is ultimately driven by changes in incident solar radiation because of long-term changes in the Earth´s orbit around the Sun. There are several climate models based on Milankovitch cycles, usually calibrated with known climatic data from previous glaciation periods. These models thus allow forecasts to be made of the future climate. For SKB/TVO's ice age scenario, the ACLIN (Kukla, 1979) and Imbrie & Imbrie (1980) models were chosen. Both models show that conditions similar to those of the present warm period will recur in 120 000 years, although a period with a relatively warm climate can, however, be expected in 75 000 years. Since this is the first time when human habitation in the present sense could again be expected in Scandinavia, after a long period of glaciation, this is also the time to which SKB/TVO's ice age scenario extends. The scenario describes climatic conditions for Scandinavia as a whole, and specifically for the Stockholm-Helsinki region (see Figure 1).

Further refinement of the dynamics of future glaciation have taken place with the development of a time-dependent model of the glaciation in Scandinavia. The model was calibrated using evidence of change in erosion/deposition and relative sea level during the last glacial cycle (Boulton & Payne, 1992).

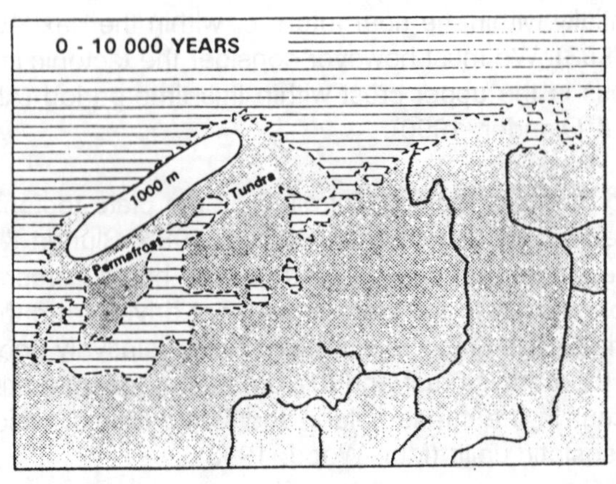

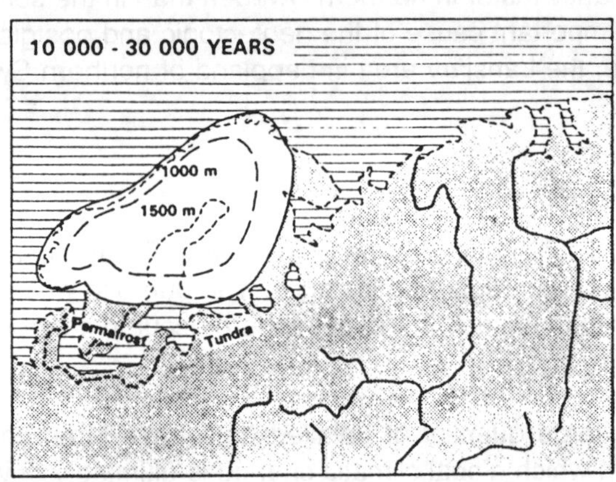

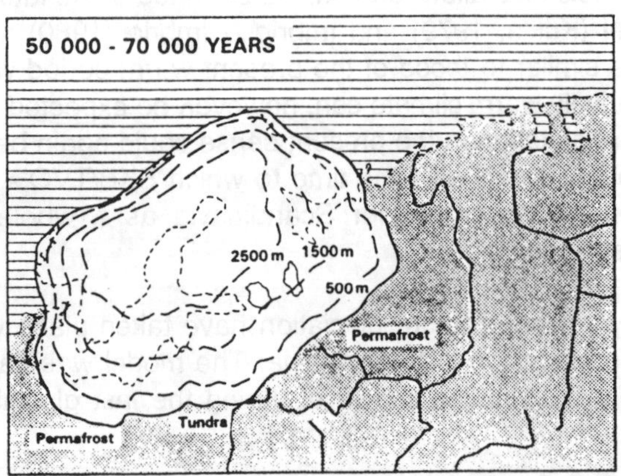

Figure 1 Extension of future ice sheet according to SKB/TVO's ice age scenario (modified after Ahlbom et al., 1991).

MECHANICAL STABILITY

Neotectonics, Isostatic Land Uplift, Postglacial Movements

"Neotectonic movements" are those which have taken place or are taking place during the current tectonic regime, i.e. during the geological period when the Atlantic Ocean has existed (about 60 million years). The last glacial period ended in Scandinavia between 12000 and 8000 years before present. Land uplift took place as a result of unloading as the ice sheet decayed. It is possible to reconstruct land uplift by reconstructing past relative sea levels from marine and lake shorelines and sediment cores. SKB has initiated studies on several sites in Sweden in order to obtain better knowledge of the pattern and rate of land uplift (Påsse, 1990). Spatial discontinuities in the patterns of postglacial land uplift may reflect tectonic discontinuities.

During the period 1986-1988, SKB carried out a research programme in the Lansjärv area, approximately 150 km north of Luleå, to attempt to map the extent and character of the presumed postglacial fault displacements in the area. In 1990, field work was concentrated on the Molberg Fault and test pits were dug in the Lansjärv area to seek for seismites, structures in the unlithified sediments which are presumed to have been formed in conjunction with earthquakes.

In June 1991, SKB arranged an excursion with international participation in the Lansjärv area, to present in the field the most important results of the Lansjärv investigations and to discuss problems of postglacial tectonics. Some conclusions of the excursion were as follows (Stanfors & Ericsson, 1993):

- Postglacial faulting occurs primarily by reactivation along older fault zones, but some new fracturing cannot be ruled out.

- The causes of the postglacial movements are probably a combination of relatively rapid changes in the vertical loads (earthquakes associated with deglaciation) and horizontal compression from the Mid-Atlantic Ridge related to continental drift. Earthquakes and seismites are clear indications of temporary instability.

Rock Mass Response to Glaciation

Future glaciation scenarios permit us to assess loading changes at the upper boundary of a rock mass. Rock mechanical modelling then permits evaluation of the potential impact. Initially, the instantaneous mechanical response of a rock mass to glaciation, deglaciation and water pressure from an ice lake have been analysed and a sensitivity study been carried out for the Finnsjön study site, Sweden (Rosengren & Stephansson, 1990; Israelsson et al., 1992). The results suggest that the strain will be taken up by pre-existing fractures. The total peak relative movement in a zone within the area amounts in the normal case to about 0.05 m. If an extreme situation prevails with low in-situ stresses in the rock, the importance of the pore pressure would increase, resulting in a total movement of about 0.5m. Current repository concepts within SKB require that canisters will not be deposited in such zones.

MODELLING OF FUTURE HYDROGEOLOGICAL CONDITIONS, INCLUDING EFFECTS OF GLACIATION

The climate and environment will change in the future in response to the natural evolution of the earth system and/or human-induced changes. Anthropogenic effects might delay a transition to glacial conditions, but it is believed that the world's climate will be dominated by glaciations over the next

100 000 years in the same way as it has been over similar periods for the past million years.

In evaluating the consequences of global change for a particular waste repository, the correlation between records of past global change and past local climate records can be investigated with the aim of predicting future local changes from the prediction of future global changes.

The purpose of an ongoing programme at SKB is:

- to identify the principal climatically-driven processes that, over a time-scale of 100 000 years, could affect the integrity of a deep waste disposal site and influence the dispersal of radionuclides from it,
- to develop models of these processes that can be first constrained by and then tested against the geological record of these processes in the past, and
- to develop a future climate function that can be used to drive the process models and produce a probabilistic estimation of the future operation of these processes and potential impacts on a specific repository site.

The processes most likely to be of importance for an underground waste repository and the dispersal from it of nuclides to the biosphere are fracturing due to loading by glaciers or freezing by permafrost, and meltwater flow beneath both glaciers and permafrost.

A time-dependent, thermo-mechanically coupled, three-dimensional model of ice sheet behaviour developed by Boulton & Payne (1992; 1994) has been used to investigate these processes. The model is driven by changes in the elevation of the permanent snowline on its surface and by air temperature and predicts the behaviour of the ice sheet for an earth's surface of given form and mechanical properties. Figure 2 shows the inputs and outputs of the ice sheet model.

A local climatological record for NW Europe has been derived from reconstructions of sea-surface temperature (SST) in the NE Atlantic, derived from studies of the geology of ocean cores. A relationship was established between the NE Atlantic sea level air temperatures and the equilibrium line altitude (ELA) required to drive the ice sheet in the way that geological evidence suggests it varied through the last glacial cycle. The time-dependent behaviour of the ice sheet model was examined along a 1620 km flowline running from the continental shelf of Norway to central Poland, and along it, ice sheet thickness, ice sheet temperature distribution, basal temperature, basal melting rate and velocity distribution were reconstructed.

Some model tests were carried out in order to establish the sensitivity of the basal temperature/melting distribution, and the distribution of permafrost below and beyond the ice sheet to variations in the assumptions used about surface temperature. The model typically predicts a very broad zone of basal melting between a narrow terminal frozen zone, a few tens of kilometres in width, and a zone of freezing beneath the ice divide.

The model was constrained by the presumed time/distance fluctuation of the European Weichselian ice sheet along the investigated transect. Testing of the model was undertaken by comparing theoretical output from the model with geological data. Three different sorts of data were selected for this test programme, the areal pattern of expansion and contraction of the ice sheet, relative sea level changes and the pattern of till distribution. It was concluded that model output was generally compatible with large parts of the geological evidence.

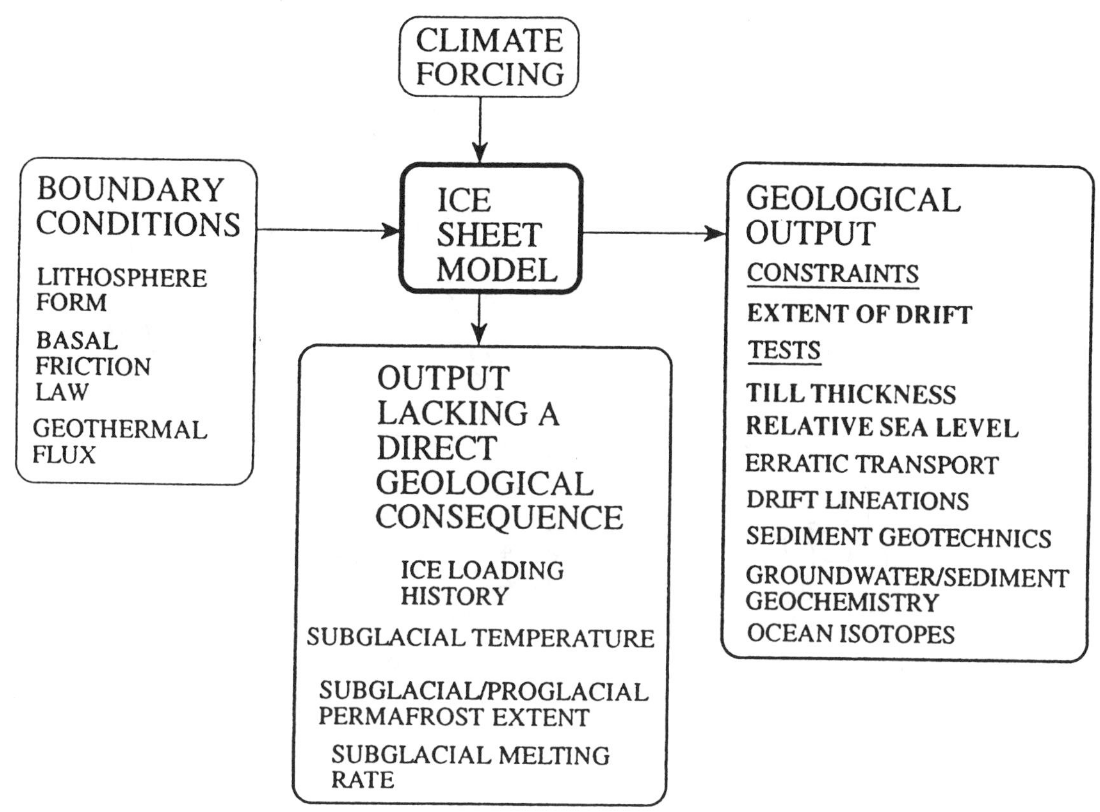

Figure 2 Inputs and outputs of the ice sheet model (Boulton & Payne, 1992).

The model was then used to give predictions of future ice sheet variations in Europe (see Figure 3), including basal temperatures, melting rates and permafrost thickness etc. Estimates of the probability of sites at different distances from the initial ice divide being glaciated at given future times were also presented.

The next phase of the research programme was a parameter study involving a number of sensitivity tests designed to establish critical parameters on which further modelling should concentrate. Wallroth and Gustafson (1993) provided empirical input and validation data from a selected broad-path transect running from the northern part of Dalarna County to Äspö island. The data collected comprised topography, lithology and drift, soil cover, groundwater chemistry, geothermal properties, rock mechanical data and lake statistics. Wallroth and Gustafson (1993) also proposed a general simplified conceptual model for the depth-dependence of hydraulic conductivity as well as quantitative models for the different lithological units along the transect on basis of available hydraulic test data.

Boulton et al. (1993) modelled the system by coupling the glaciation model to a model of subglacial Darcian groundwater flow. A major purpose of the work was to evaluate the sensitivity of the subglacial melting rate to a series of possible controls. These were glacier dynamics, variations in geothermal flux and topography. The transect was further extended and deflected in a southwesterly direction to link with a transect across the part of northern Germany and Denmark where the Quaternary and Mesozoic aquifers develop (Boulton et al., 1994).

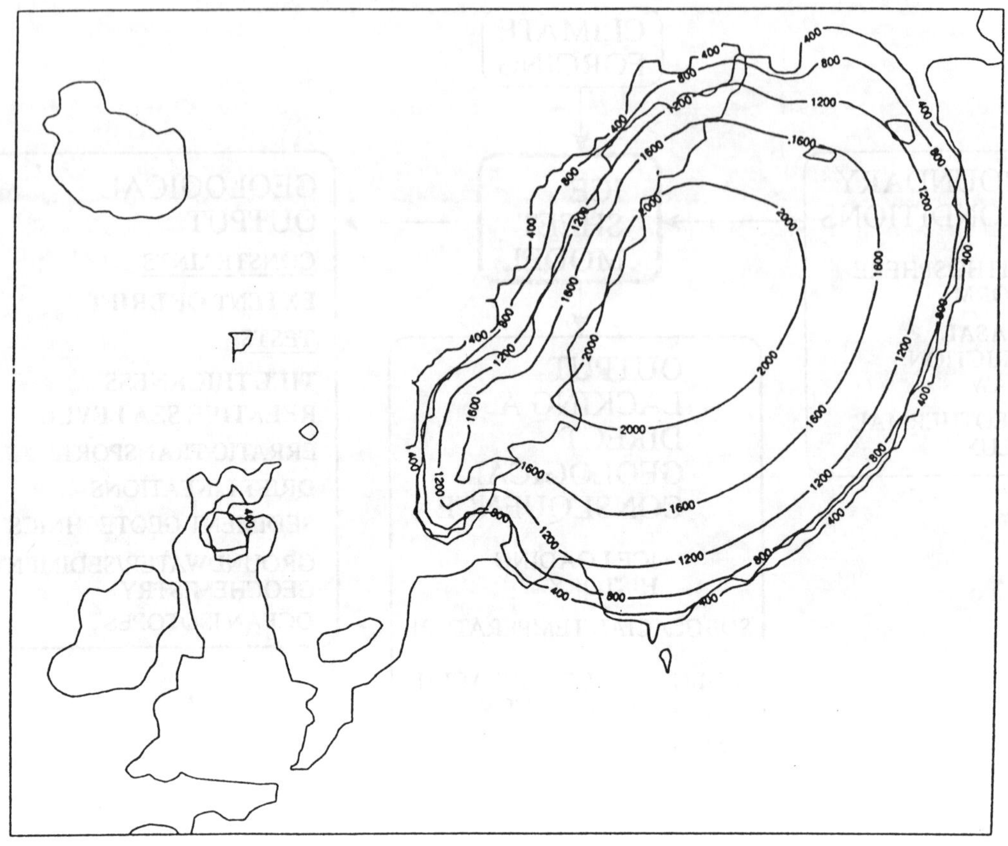

Figure 3 Predicted extent of the ice sheet over northern Europe at the next glacial maximum in about 70 000 years time (Boulton & Payne, 1992).

Changes in ice sheet dynamics strongly influenced the rate of basal melting by changing the thermal gradient in the upper lithosphere or ice and by frictional heating at the base of the glacier. A number of parameters; basal friction, sea level temperatures, vertical temperature gradient in the atmosphere, and accumulation rate on the ice sheet surface were varied in order to investigate the relationships. From these computations it was concluded that basal melting rate was sensitive to changes in the atmospheric lapse rate and to changes in the accumulation rate.

The rate of basal melting was found to be relatively insensitive to variations in the geothermal flux and topographic details. However, the results showed that dynamic characteristics that indirectly influence the basal melt rate, such as spacing and size of valleys, had an important effect. Large inter-valley spacings and deep valleys generate larger melt rates.

Validating the Model

The results of these sensitivity tests indicated possible ranges of basal melting rates and subglacial water discharges. According to these calculations melt rates at the base of an ice sheet are likely to fall within the range 2-100 mm/year. If the transmissivity of the bed is inadequate to discharge by groundwater flow alone the total subglacial melt discharge, tunnels will form at the ice/bed interface and these will transport a significant part of the total meltwater flux. On the other hand, if the transmissivity of the bed is very large, it may have the capacity to discharge all the melt by groundwater flow and there will be no tunnel flow. Subglacial discharges, maximum subglacial potential gradients and maximum groundwater discharge were computed for the given range of melt rates and for ice sheets of varying span. By subtracting the groundwater flux from the total melt flux a prediction was produced of the tunnel discharge at the glacier bed. Over the relatively impermeable bedrock of much of Sweden, tunnel flow will dominate the discharge, but over the thick permeable aquifers of Scania, Denmark, Germany and Holland all the melt flux can be accommodated by groundwater flow.

A means of testing (validating) the melting model was then identified. Eskers, long ridges of sand and gravel, are the sedimentary expression of subglacial tunnels. If, along a glacier flowline, bed transmissivity increases successively from very small to very large values, tunnels (and eskers) will cease at a point where the capacity of groundwater to discharge the meltwater flux is achieved. Such a point is therefore a measure of the melt flux (Boulton et al., in press). The frequency and average separation of eskers along three different transects through the Fennoscandian ice sheet were measured and compared with the estimated transmissivity of the bed. It was found that esker distributions were compatible with a discharge value corresponding to an average melting rate of between 5 and 30 mm/year. By applying this rate to the subglacial groundwater model, it has been estimated that groundwater from 500 m depth may be driven to the surface over periods of the order of 10^3-10^5 years.

PALAEOHYDROGEOLOGICAL MODELLING DEVELOPMENT

The next phase of the research programme has the objectives of simplifying the ice sheet model and its coupling to groundwater flow, investigating glacially-generated fracturing of bedrock and generating boundary conditions for a detailed study of time-dependent groundwater flow in the Äspö-Laxemar region.

A further test of the climate/ice sheet model and an understanding of bedrock conductivity will be based on the chemical variability of groundwater recharge, depending upon climate, and the isotopic composition of recharge from glacial meltwater. The climate/ice sheet model is able to simulate the isotopic variability of recharge water, and we now wish to establish the distribution in groundwater of the geochemical signal of past climate change as a means of determining long-term rock conductivity.

In order to investigate the regional hydrogeological and hydrogeochemical conditions, a deep borehole (1700 m) has been drilled in the Laxemar area near the Simpevarp peninsula and the Äspö HRL. The coredrilling is the deepest one in Scandinavia in crystalline rocks. After concluded drilling, geophysical, hydrochemical, geochemical and hydraulic investigations have been performed in the hole and rockmechanical investigations are planned.

The regional simulation of groundwater flow will be transient, i.e. through the whole of the last interglacial and glacial cycle. An essential part of the model testing and validation will be allocated to the hydrochemistry results from the Äspö HRL and Laxemar projects. Special interest is devoted to the investigations of stable (O, H, S, C) and radiogenic (Sr) isotopes of groundwater and calcite fracture fillings. The integration of these isotopic techniques, along with conventional water quality data, provides a powerful approach for understanding the origin and evolution of the groundwater and of the hydrogenic minerals.

REFERENCES

Ahlbom, K, Äikäs, T and Ericsson, L O, 1991: SKB/TVO ice age scenario. SKB Technical Report 91-32, Svensk Kärnbränslehantering AB, Stockholm.

Björck, S and Svensson, N-O, 1992: Climatic changes and uplift patterns- past, present and future. SKB Technical Report 92-38, Svensk Kärnbränslehantering AB, Stockholm.

Boulton, G S, 1991: Proposed approach to time-dependent or "event-scenario" modelling of future glaciation in Sweden. SKB Arbetsrapport 91-27, Stockholm.

Boulton, G S and Payne, A, 1992: Simulation of the European ice sheet through the last glacial cycle and prediction of future glaciation. SKB TR 93-14, Stockholm.

Boulton, G S, Caban, P and Punkari, M, 1993: Sub-surface conditions in Sweden produced by future climate changes, including glaciation. Project 2, Sensitivity tests and model testing. University of Edinburgh, Edinburgh, UK (draft report to SKB).

Boulton, G S, Slot, T, Blessing, G, Glasbergen, P, Leijnse, T and van Gijssel, K, 1994: Deep circulation of groundwater in overpressured subglacial aquifers and its geological consequences. Quaternary Science Reviews, 12, 739-745.

Boulton, G S and Payne, A, 1994: Mid-latitude ice sheets through the last glacial cycle: glaciological and geological reconstructions. In: Duplossy, J-C and Spyridakis, M-T, Long-term climatic variations. NATO ASI Series, Vol. 122, Springer-Verlag, Berlin.

Boulton, G S, Punkari, M, Caban, P E and Wallroth, T, in press: Eskers, subglacial groundwater flow and deduction of subglacial melting rates and water discharges from former ice sheets. Journal of Glaciology.

Eronen, M and Olander, H, 1990: On the world's ice ages and changing environmens. Report YJT-90-13, Nuclear Waste Commission of Finnish Power Companies, Helsinki.

Imbrie, J and Imbrie, J Z, 1980: Modelling the climatic response to orbital variations. Science, 207, 943-953.

Israelsson, J, Rosengren, L and Stephansson, O, 1992: Sensitivity study of rock mass response to glaciation at Finnsjön, central Sweden. SKB Technical Report 92-34, Svensk Kärnbränslehantering AB, Stockholm.

Kukla, G, 1979: Probability of expected climatic stresses in North America in the next one M.Y. In: Scott, Craig, Benson and Harwell (eds). A summary of FY-1978 consultant input for scenario methodology development. Pacific Northwest Laboratory of Battelle Memorial Inst. PNL-2851.

Larson, S Å and Tullborg, E-L, 1993: Tectonic regimes in the Baltic Shield during the last 1200 Ma-A review. SKB Technical Report 94-05, Svensk Kärnbränslehantering AB, Stockholm.

Milankovitch, M, 1941: Kanon der Erdbestrahlung und seine Anwendung auf das Eiszeitproblem. Royal Serbian Sciences, Spec. Publ. 132.

Muir-Wood, R, 1993: A review of the seismotectonics of Sweden. SKB Technical Report 93-13, Svensk Kärnbränslehantering AB, Stockholm.

Påsse, T, 1990: Empirical estimation of isostatic uplift using the lake-tilting method at Lake Fegen and Lake Säven, southwestern Sweden. Mathematical Geology, Vol. 22, No. 7.

Rosengren, L and Stephansson, O, 1990: Distinct element modelling of the rock mass response to glaciation at Finnsjön, central Sweden. SKB Technical Report 90-40, Svensk Kärnbränslehantering AB, Stockholm.

Slunga, R, 1985: The seismicity of southern Sweden, 1979-1984, final report. FOA Report C 20572-T1, Försvarets Forskningsanstalt, Stockholm.

Slunga, R and Nordgren, L, 1987: Earthquake measurements in southern Sweden, Oct. 1 1986-March 31 1987, SKB Technical Report 87-27, Svensk Kärnbränslehantering AB, Stockholm.

Slunga, R, 1989: Earthquake mechanisms in northern Sweden, Oct. 1987- April 1988. SKB Technical Report 89-28, Svensk Kärnbränslehantering AB, Stockholm.

Stanfors, R and Ericsson, L O (eds), 1993: Post-glacial faulting in the Lansjärv area, northern Sweden. Comments from the expert group on a field visit at the Molberget post-glacial fault area, 1991. SKB Technical Report 93-11, Svensk Kärnbränslehantering AB, Stockholm.

Wallroth, T and Gustafson, G, 1993: Sub-surface conditions produced by future climate changes, including glaciation. Data support for modelling. SKB Arbetsrapport 92-77, Stockholm.

The Central Climate Change Scenario: SKI's SITE-94 Project to Evaluate the Future Behaviour of a Deep Repository for Spent-Fuel

Louisa M King, Neil A Chapman
Intera Information Technologies, 47, Burton Street, Melton Mowbray, Leics., UK
Fritz Kautsky
Swedish Nuclear Power Inspectorate, P.O. Box 27106, S-102 52, Stockholm, Sweden

Abstract

Part of the SITE-94 project currently being performed by the Swedish Nuclear Power Inspectorate (SKI) involves the construction of scenarios to assist in the evaluation of the future behaviour of a deep repository for spent-fuel. The project uses real data from the Äspö site, and assumes that a hypothetical repository (reduced in size to approximately 10%) is situated at about 500m depth in granitic bedrock in this coastal area of SE Sweden. A Central Scenario, involving prediction of the climate and consequent surface and subsurface environments at the site for the next c.120,000 years, lies at the heart of the scenario definition work. The Central Climate Change Scenario has been based on the climate models ACLIN (Astronomical climate index), Imbrie & Imbrie (1980) and the PCM model by Berger et al. (1989). These models suggest glacial maxima at c. 5000, 20,000, 60,000 and 100,000 years from now. The Äspö region is predicted to be significantly affected by the latter three glacial episodes, with the ice sheet reaching and covering the area during the latter two episodes (by up to c. 2200m and 1200m thickness of ice respectively).The objective of this work is to provide a first indicator of the physical and hydrogeological conditions below and at the front of the advancing and retreating ice sheets, with the aim of identifying critical aspects for modelling impacts of future glaciations on far-field groundwater flow, rock stress and groundwater chemistry. The output of this study is a time-dependent description of the properties of the site, coupled to predictions of key parameter values at different times in the future, for input into performance assessment modelling.

1. Introduction

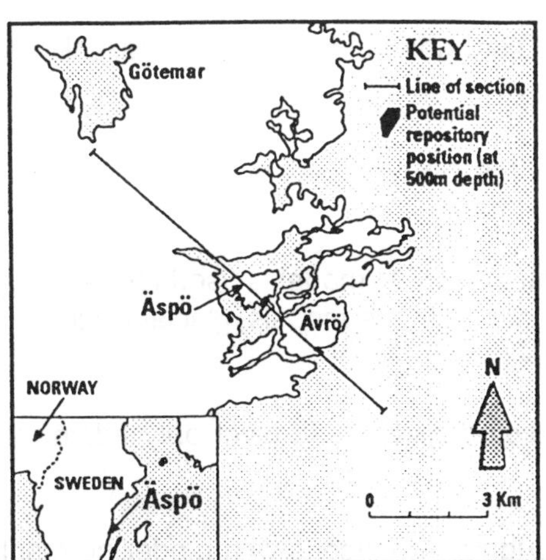

Part of the SITE 94 Project involves the construction of scenarios to assist in the evaluation of the future behaviour of a deep repository for spent-fuel. The project uses data from the Äspö site and assumes that a repository (reduced in size to approximately 10%) is situated at about 500m depth in granitic bedrock in this coastal area of SE Sweden (Figure 1). A Central Scenario, involving prediction of the climate and consequent surface and subsurface environments at the site for the next c. 120,000 years, lies at the heart of the scenario definition work.

Figure 1. **The location of Äspö with the line of section used to construct the Central Scenario**

The scenario construction process first involves the definition of all the features, events and processes (FEPs) which affect the behaviour of a repository system. The Central Scenario thus requires the following main components: a deterministic description of the most probable future climate states for Sweden, with special reference to the Äspö area; a description of the likely nature of the surface environment in the site area at each stage of the climate sequence selected; quantitative information on how these changes might affect the disposal system for input in to performance assessment modelling. As recognised by Chapman et al (in press), we see the Central Scenario as simply a means of *illustrating* possible future behaviour of the system and exploring how such behaviour might arise.

2. On the relation between past and future climates

Predicting future climate changes is a difficult task requiring knowledge and understanding of the earth's past climate changes, as well as requiring a model for predicting the future. It is now known that a period of glaciations of the Northern Hemisphere started some 2.3 million years ago, and since then, during the Quaternary period, a considerable number of glaciations, interrupted by warmer interglacials, have been documented. The periodicity of the glacial/interglacial sequence approaches a 100,000 year cycle during the middle and late Pleistocene. This knowledge of Quaternary glaciations allows us to make a preliminary prognosis of future glaciations. The last interglacial, the Eemian (c. 120,000-135,000 yrs BP), was c.10,000-15,000 years long. As the present warm period, the Holocene, has lasted about this long, and as the Holocene climatic optimum passed several thousand years ago, it is possible that the cooling of the next glaciation has already begun. The great uncertainty in this simple prediction is that it does not consider the causes of the glacial/interglacial cycles. A more elaborate prognosis of future climate would need a climate model based on the variables which regulates the glacial/interglacial cycles.

The present paper aims to present a prognosis of future climate and glaciations based on a number of published climate models; ACLIN (Astronomical climate index), the Imbrie & Imbrie (1980) model and the PCM model by Berger et al. (1989). All these models consider the Milankovitch orbital parameters. From these models changes in the global amount of ice and corresponding eustatic sea-level were estimated (Figure 2). The prediction of coming glaciations in Fennoscandia was based on parallels with the last glaciation. In the following sections important factors contributing to the scenario will be discussed, as well as those significant factors not taken into account.

2.1 Variables and assumptions contributing to the climate scenario construction

- The presented scenario is based on orbitally induced changes in insolation, the Milankovitch forcing.
- For future glaciations an ice model with a single Fennoscandian dome and maximum ice thickness of 3000 m is used for a major glaciation.
- The ice volume calculated in the Imbrie & Imbrie model and PCM model is used for estimating future global sea-level fluctuation. For periods not included in these models, glacial volume and global sea-level are subjectively derived from the ACLIN model.
- The local sea-level is derived directly from the predicted global sea-level only adding the predicted glacio-isostasy.
- The isostasy during glacial episodes is estimated through comparisons with modelling results for the last Weichselian glaciation given by Fjeldskaar (unpublished data, and Fjeldskaar & Cathles 1991). This model applies a low viscosity mantle (1.6 x 10^{22} poise), a 75-100 km thick astenosphere (0.7 x 10^{20} poise) and a lithosphere close to 90 km thick.

- The distribution and extent of ice during forthcoming glaciations is derived by direct parallels with the Weichselian. The timing of future Scandinavian glaciations is assumed to be in phase with the global ice volumes of the models used.

2.2 Important variables not accounted for in the scenario construction

- To explain the dominance of the 100,000 year cycles some other factors such as varying response times, CO_2 effects, glacio isostatic downwarping, basal sliding of the ice sheets and the feedback between sea-level and ice-sheets, have to be included in the models. To some extent this is done in the models used but more work is needed.

- The global glacial volume as predicted in the models, and thus global eustasy, need not be in phase with the Scandinavian ice buildup. Thus the local sea-level changes could be difficult to predict.

- The varying estimates of thickness of the Weichselian ice-sheet could give large discrepancies in modelling glacio-isostasy as well as in making parallels for future glaciations. The ice model applied in the scenario, a thick single Fennoscandian ice dome, could have alternatives such as a much thinner ice or an ice sheet with separated domes. The last glaciation to affect Scandinavia may have involved the formation of several coalescing domes (Anundsen, 1993).

- In predicting the response of earth crust during future glaciations and sea-level fluctuations, detailed geophysical modelling would have been of great value. The effects of the forebulge and hydro-isostasy could be accounted in predicting sea-level changes and glacial isostasy could be determined more accurately.

- The CO_2 content of the atmosphere seems to be an very important moderator on climate changes. According to ongoing modelling by Berger (pers. communication, 1994) the CO_2 content may imply that the predicted onset of a major glaciation at 65°N north will first start c. 55 K AP and quickly go into a glaciation phase.

- "Flip-flop" variation of climate. The new results from central Greenland ice cores show that the climate during the last 230,000 years has been unstable, with abrupt climate changes. These findings further emphasise that the Milankovitch forcing only gives parts of the background for climatic changes.

- Human alteration of climate development through effects by human activities (e.g. greenhouse warming) have not been considered within the Central Scenario, as they are difficult to assess quantitatively in detail. Human effects could prolong the present state of climate or result in global warming, but this effect is not considered to hinder the next cold phase, merely delay it in the perspective of the next coming 120,000 years. However, an alternative, human induced warmer climate scenario is considered within the SITE 94 scenario exercise (Chapman et al., in press).

3. The SITE 94 Central Scenario - The Climate Change Scenario

The present scenario aims to provide an *illustration* of future climatic change in SE Sweden. The predictions of how the climate will change over the next 130,000 years is, however, limited by substantial uncertainties (compared to the actual possibly measurable future evolution of the climate). It is obviously impossible to give true quantitative values for any parameter during the seven different hypothetical evolution stages described below. The presented estimations should only be looked upon as rough measures and should serve as comparisons between the stages.

The climate models used for this study suggest glacial maxima at c. 5000, 20,000, 60,000 and 100,000 years from now. The scenario is based on earlier scenarios (Ahlbom et al. 1991; Björck & Svensson, 1992) with some modifications due to input from McEwen & de Marsily (1991), Boulton & Payne (1992) and Svensson (1989) and King & Chapman (1994). The proposed climate evolution is illustrated in Figure 2. The Äspö region is predicted to be significantly affected by the latter three glacial episodes, with the ice sheet reaching and covering the area during the latter two episodes (by up to c. 2200m and 1200m thickness of ice respectively). Only periglacial conditions are predicted to affect the area during the glacial episode at c. 20,000 years from now and no glacial effects are predicted at c. 5000 years. Permafrost development has been modelled (King et al., in prep), elaborating on work by McEwen & de Marsily (1991) and based on temperature data from Boulton & Payne (1992). The results are shown on Figure 2 (see King et al, in prep, for details).

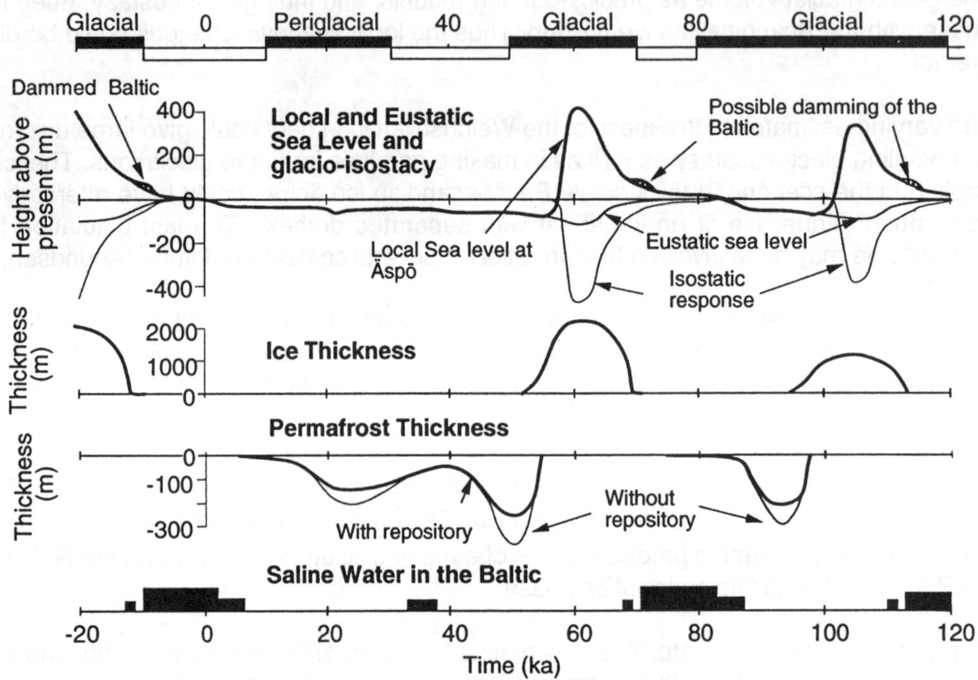

Figure 2. **Predictions and observations of global and local (Äspö) eustatic sea-level; sea-level variations, ice cover, glacio-isostasy, permafrost thickness and Baltic salinity. Past eustatic record from Fairbanks (1989) and past shore-displacement from Svensson (1991)**

0 - 10 000 years. The climate in Sweden will gradually change to cooler conditions with growth of an ice sheet in the Scandinavian Caledonides. Ice sheet thickness in the central mountains c. 1000 m - no ice in Stockholm or southwards. Crustal downwarping of c. 300 m in the central mountainous part. Sea level will gradually drop to -20 to -40 m below present day sea-level in the Stockholm region as well as in the Äspö region. During the colder parts of this period permafrost will occur in northern Sweden. The water in the Baltic will gradually become fresher as the ocean connection decreases.

10 000 - 30 000 years. After a minor, somewhat warmer, period the climate will get colder and fully glacial conditions will prevail around 20,000 years from now. The glacial peak will last perhaps 5000 years. The ice sheet is estimated to reach the Stockholm area but probably not the Äspö region. Ice thickness in the central part of the ice sheet will be c. 1500 m, while the ice sheet thickness in the Stockholm region will be c. 800 m. Crustal downwarping of about 500 m in the central part of the ice sheet and c. 60 m in the Stockholm latitude. During deglaciation, when the ice front is located at the Stockholm region, the sea-level is estimated to be c. 25 m below present day sea-level. At Äspö sea-level will drop to c. -50 m below present day sea-level. The total effect of the glacial loading at Äspö is difficult to evaluate, but will probably not be large. Permafrost conditions will existdown to southern Sweden. The water in the Baltic will be fresh, but probably in level with the oceans, due to erosion at its outlets.

30 000 - 50 000 years. Interstadial with a dry and cold climate (like the climate today on Greenland). Glaciers in the Swedish mountains and permafrost in northern Sweden. In the Stockholm region the eustatic rise of sea-level and decreasing isostatic uplift will result in a sea-level c. 30-40 m below present day sea-level. At Äspö there will not be much isostatic uplift but a eustatic sea-level rise of maybe 10-20 m may occur. Thus the sea-level will be c. 30-40 m below present day level. The Baltic will mainly be fresh but some saline water may enter it.

50 000 - 70 000 years. Full glacial conditions will prevail. Due to the previous cold conditions the ice sheet will respond more rapidly. The glacial culmination will be around 60,000 years. The ice sheet will cover the whole of Sweden down to north Germany, comparable to the maximum of the Weichselian glaciation. Ice sheet thickness in the central part will be c.3000 m. In the Stockholm region the ice thickness will be c.2500 m. The Stockholm region will probably be covered by the ice sheet for at least 10 000 years, and possibly longer. Downwarping of c.700 m will occur in the central part of the ice sheet, c.600 m in the Stockholm area and c.500 m at Äspö. During deglaciation, when the ice front is located at the Stockholm region, the sea-level is estimated to be c.150 m at Stockholm and at the same time c.80 m above present day coastline at Äspö. There could also be a damming of the Baltic to 10-30 m above ocean level. The Baltic is mainly fresh but during deglaciation saline water may enter the Baltic through isostatically depressed areas. Permafrost will be present in large areas of Europe.

70 000 - 80 000 years. A rapid deglaciation will lead to and culminate in interglacial conditions at 75,000 years. Total crustal uplift is estimated to be c.700 m in the central parts of the previous ice sheet, c.600 m at Stockholm and 500 m at Äspö. This will be a relatively "warm" period with a climate in the Stockholm region similar to the present climate in northern Sweden. Small mountain glaciers and permafrost will occur in the very north. Parts of southern Sweden will be resettled and farming might be possible. Sea-level and salinity similar to present day all over and the land uplift will restore the land-surface to approximately its present state. Permafrost will only be present in the very north of Scandinavia.

80 000 - 120 000 years. The climate will gradually become colder with maximum glacial conditions at 100,000 years. The ice sheet will be extensive, covering large parts of Fennoscandia. Permafrost will occur in large areas outside the ice-margin. In the Stockholm region maximum ice thickness is estimated to be c.1500 m, and at Äspö c.1000 m. Downwarping of the ground with c.500 m at Stockholm during the main phase of the glaciation. At Äspö the maximum downwarping will be a little less than at Stockholm, maybe 400 m. The relative sea-level at Stockholm at deglaciation will be c.100 m above present day coastline. The sea-level at Äspö at deglaciation may be c.80 m above present day sea-level, but damming of the Baltic to 10-30 m above ocean-level is possible. The Baltic is mainly fresh but during deglaciation saline water may enter the Baltic through isostatically depressed areas.

120 000 - 130 000 years. Interglacial. The next warm period with a climate similar to the present in the whole of Scandinavia. Sea-level and salinity in the Baltic similar to the present.

4. Glacial Cycle Scenario Conceptualisation

A sequence of pictorial scenarios has been constructed for a single glacial cycle (advance and retreat of an ice sheet), based on the glaciation which is predicted to climax at c. 60,000 years from now (Scenarios 1-7; Figure 3).

The intention is that the scenarios represent the most significant features, events and processes for analysis of the future safety of a repository for spent fuel. Thus, the objective of this work is to provide a

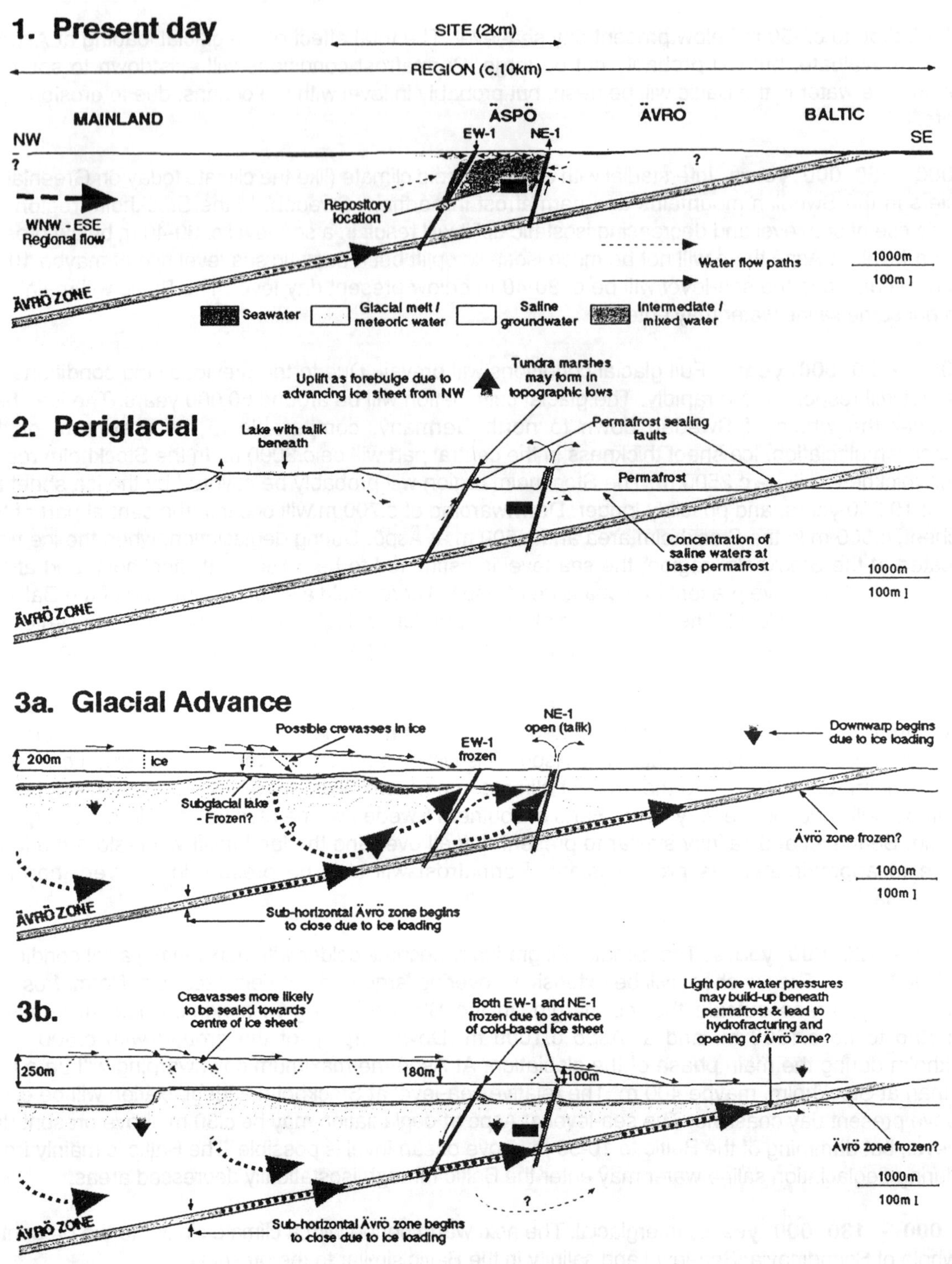

Figure 3. **Scenarios for the Äspö area and region through the single glacial cycle predicted to occur from 50,000 to 70,000 years from now.**

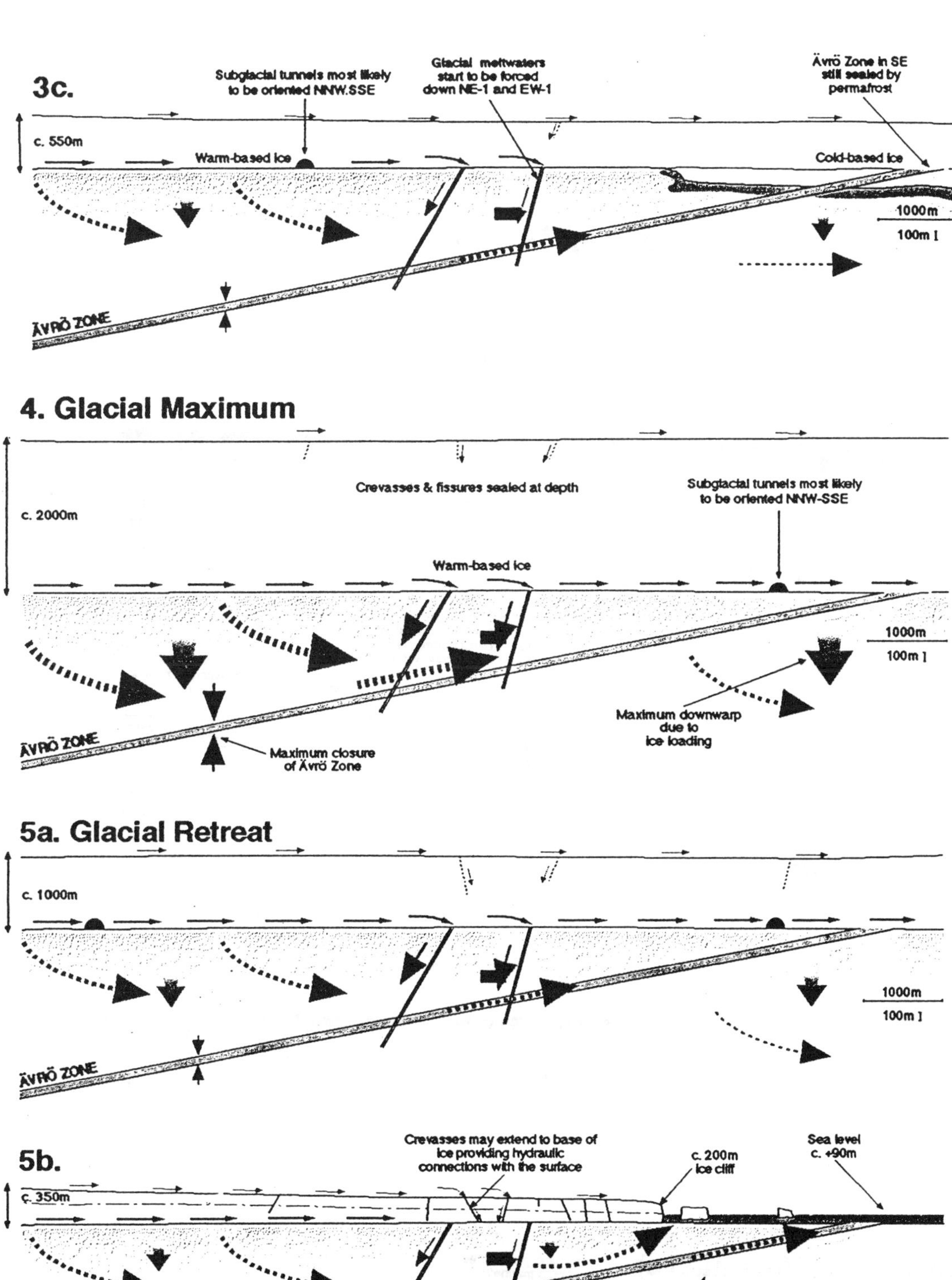

Figure 3. **(Continued)**

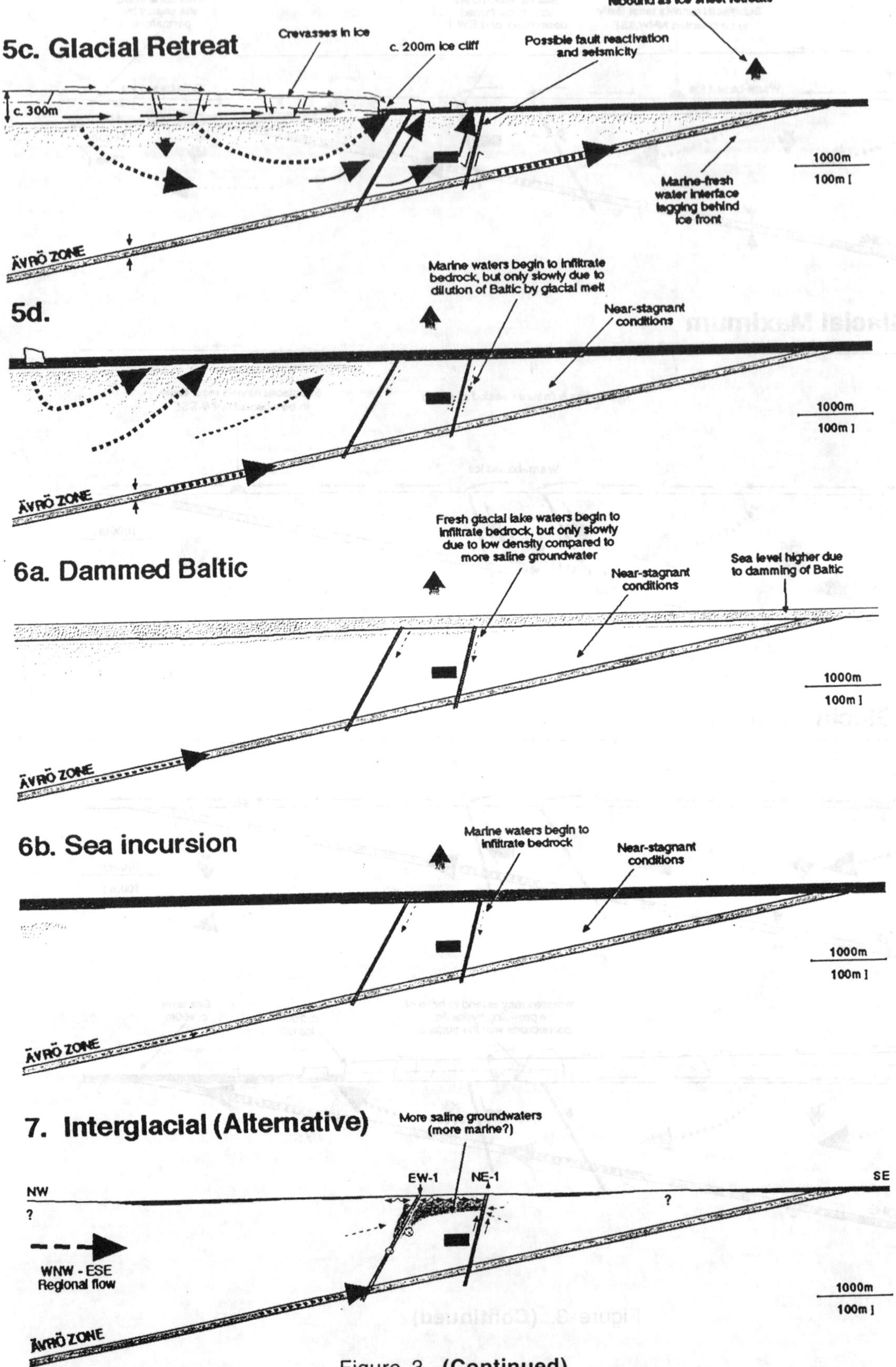

Figure 3. **(Continued)**

first indicator of the physical and hydrogeological conditions at the front of and beneath the advancing and retreating ice sheets. We have looked specifically for information on the existence of free water under the ice, hydraulic connection between free water in, on and under the ice with groundwater in the underlying rock, the rate of ice build-up and decay at a site-scale, and evidence for the geometry of the ice front during both advance and retreat. These features, in addition to other related features, such as the development of permafrost and sea level changes, are considered in terms of their impact on hydrogeology, groundwater chemistry, rock stress and surface environments for input into performance assessment modelling.

Many simplifying assumptions have been made in the development of the scenario sequence, in addition to those outlined above. The topography of the Äspö region and area is assumed to be unchanging throughout the glacial cycle. Only three fault zones (EW-1, NE-1 and the Ävrö Zone) are considered in detail. The flow paths shown within the bedrock are therefore only intended to display approximate flow directions. The tectonic setting of Äspö is considered to remain constant through the glacial cycle, such that only glacial loading or glacio-isostatic effects are considered to change. The morphology of the ice sheet assumes that it develops over a predominantly non-deformable substrate during the early and later periods of the glacial cycle, over the majority of Norway and Sweden, but that it may advance and retreat over peripheral more deformable substrate, in inland Europe and the Gulf of Bothnia, during the glacial maxima and perhaps retreat (e.g. Boulton & Jones, 1979; Holmlund & Fastook, 1993). Surface profiles have been determined by consideration of maximum ice thickness estimates and the horizontal extent of the ice sheet, modelling by Boulton & Payne (1992) and by comparison with the Antarctic and Greenland ice sheets (e.g. Skinner & Porter, 1987; Embleton & King, 1975). Repetition of phases, such as multiple advance and retreat due to ice surging or climatic changes within the glacial cycle, in addition to the formation of ice lobes and interlobate complexes, have not been considered.

5. Glacial Cycle Scenarios

5.1 Present day / Interglacial (Scenario 1)

The base case scenario is that of an interglacial period with relatively warm conditions free from the influence of permafrost or ice, akin to those at the present day at Äspö. The areal and regional scenarios are depicted in Scenario 1 (Figure 3) and have been modified from Smellie & Laaksoharju (1992). The line of section is orientated approximately NW-SE (Figure 1). It is assumed that the hypothetical repository for spent fuel is situated at a depth of 500 m, in the SE of Äspö and that it has been closed just prior to this scenario. For the purposes of the Site 94 modelling, the repository is considered to be approximately one tenth of the size of an actual repository, comprising 400 canisters.

At present the regional maximum principal stress is known to be oriented in a predominantly NW-SE direction. Glacio-isostatic stresses are considered to be zero for this base case although isostatic rebound may still be occurring over Scandinavia (e.g. Anundsen, 1993).

The nature of present day groundwater flow in the Äspö area and region has been simplified from Smellie & Laaksoharju (1992) and SITE 94 work in order to accommodate modelling. Regional groundwater flow, reflecting the regional topographic gradient, is taken to be approximately WNW to ESE although more localised deep groundwater flow occurs perpendicular to the line of section (Figure 3). The various fracture zones act as zones of recharge and discharge.

The Base Case groundwater chemistry assumes the presence of five groups of water classified according to ^{18}O, D and Cl contents (P. Glynn & C. Voss, pers comm., 1994): recent Na-HCO_3-rich waters present to a few hundred metres depth; old, dilute glacial meltwater group present in isolated shallow sections in the northern part of Äspö; highly saline deep water group found at depths greater than about 500 m; intermediate 4000 to 6000 mg/l Cl group found at c. 200 to 500 m depths; group with a seawater or Baltic signature present only in very isolated areas below 300 m. It has been assumed that the combination of these chemistries resulted solely from the previous glacial cycle, although this is unlikely.

5.2 Periglacial: c.48,000 years (Scenario 2)

Permafrost is considered to be present in the Äspö region to a maximum depth of about 250m (Figure 2) although it is punctuated by taliks. NE-1 has been assumed to form a talik, whereas both EW-1 and the Ävrö Zone are assumed to be sealed by the permafrost to a depth of 250m. Isostatic uplift is considered to be negligible during this period, although development of a forebulge in front of the advancing ice sheet situated further north may affect Äspö. The forebulge may reach a maximum height of c. 80m at Äspö when it is approximately 200km in front of the ice sheet margin (e.g. Boulton, 1991; cf. Walcott, 1970 and Mörner, 1979). Sea level is approximately 50 m below the present day level at Äspö for this scenario. Lakes may develop on the periglacial surface within topographic depressions, perhaps where the sea once was present in the base case. Vegetation will be scarce in these tundra conditions and it is likely that the ground surface will be characterised by bare rock with only local organic accumulations as tundra marshes in isolated poorly-drained topographic lows which may develop along outcropping fracture zones.

During this period there will be no direct ice effects on the bedrock at Äspö. However, as the ice sheet advances towards Äspö, development of a flexural forebulge will affect stress patterns within the bedrock (e.g. Muir Wood, 1993; Shen & Stephansson, 1990). Tensional opening or reactivation of fractures may occur during forebulge uplift prior to subsequent compressive loading. However, there is apparently no evidence for reactivation of fractures during glacial advances and fracture reactivation is therefore not considered in this scenario. As permafrost develops, the formation of ice within any fractures in the bedrock, and the resulting expansion of the fracture water volume, will exert stresses that tend to propagate the fractures although they will remain sealed within the permafrost zone.

Infiltration and recharge will be severely restricted, particulary where the zones of base case flow are sealed by permafrost (EW-1 and the Ävrö Zone). If taliks completely penetrate the permafrost, pathways will be available either for recharge or discharge of sub-permafrost water. If a lake existed before development of the permafrost, an open talik could develop (McEwen & de Marsily, 1991). The concentration of recharge in smaller areas and fewer points can lead to increased recharge and discharge rates at these points. Although permafrost is not considered to form at repository depths, it may well have an indirect effect on the repository by altering the volume of groundwater flow, as well as local flow directions, through the repository. Forebulge uplift is likely to cause a re-orientation of regional flow directions, assumed here to be predominantly a function of topographic gradient, until the forebulge has passed away from the Äspö region.

The formation of very saline waters may occur at the base of the permafrost zone due to the longer residence times of groundwaters trapped beneath the permafrost (McEwen & de Marsily, 1991). If radionuclides are present in the groundwaters, they may become concentrated amongst these saline waters beneath, or within, the permafrost zone. Alternatively, they may be concentrated at the lower levels of chemically stratified lakes acting as lacustrine taliks. Subsequently, the radionuclides may be released as a concentrated pulse when thawing takes place (see below).

5.3 Glacial advance: 50,000 - 60,000 years (Scenarios 3a-c)

During the first c. 5,000 years of this phase the sea level at Äspö is predicted to be below present day sea level such that the ice sheet will be completely landbased as it advances and will initially reach Äspö on dry land (3a). The margin of the ice sheet is expected to have an approximately WSW-ENE orientation as it passes across Äspö. The ice sheet will advance across permafrosted bedrock, possibly with permafrost reaching depths up to 250 m. The ice sheet will therefore be cold-based until temperatures at the base of the ice sheet reach the pressure melting point and the permafrost melts (Hindmarsh et al., 1989; Drewry, 1986) (3c).

The flexural forebulge will prograde further ahead of the ice front and away fom the Äspö region. As the ice front comes within several tens of kilometers of Äspö the bedrock will begin to downwarp due to the loading effect of the ice sheet to the north (3a). Downwarping will continue as the ice sheet itself advances onto the Äspö bedrock (3b & c). As downwarping begins to take place, it is likely that subhorizontal fractures will begin to close due to the increasing vertical stresses. Near-vertical fractures may close due to increases in horizontal stresses, or may undergo shearing, which, in combination with increased pore water pressures, may act to enhance the conductivity of the fractures. In bulk, the bedrock will tend to experience overall compression during loading (Muir Wood, 1993).

During initial advance into the Äspö region, the ice sheet will probably be cold-based and overlie permafrost, so that little or no subglacial meltwater will be generated (Hindmarsh et al., 1989; Drewry, 1986). Groundwater recharge and discharge patterns may be similar to those under periglacial conditions (3a). However, any groundwater present within the bedrock beneath the permafrost zone underlying the ice sheet will be forced outwards and subsequently upwards in front of the ice margin due to the excess hydraulic gradients, considered, as a conservative first approximation to be controlled by the parabolic thickness profile of the ice sheet (e.g. Boulton, 1991). In addition, groundwater generated from subglacial melting further towards the centre of the ice sheet is also expected to make its way to the ice sheet margins. Regional flow beneath the ice sheet will occur in a direction approximately perpendicular to the ice margin, almost parallel to the line of section. Upward flow would therefore be expected along the fracture zones, although only NE-1 is considered to allow discharge at the surface in the scenario. Insulation by the permafrost is likely to induce the build up of high pore water pressures and hence may cause fracture reactivation and/or hydrofracturing in the outer subglacial or proglacial permafrost zones (Boulton, 1991; Muir Wood, 1989). Supraglacial meltwater will provide some surface run-off and potential recharge. As the ice sheet covers the Äspö area, recharge and discharge will be severely inhibited. Flow will be directed downwards, but is likely to remain upwards along the Ävrö Zone, although closure of the zone may start to prevent flow directly along it.

As the ice sheet advances and thickens, the base of the ice is expected to reach pressure melting point and hence melting will occur of both the permafrost and basal ice (3c). Subglacial water pressures will build up leading to subglacial water discharge, either into available transmissive zones in the bedrock, and/or along tunnels at the base of the ice where the underlying bedrock is of low transmissivity (e.g. Boulton, 1991; Clark & Walder, 1994). Ice pressures and the hydraulic gradient will tend to force the flow downwards and towards the ice front. Subglacial tunnels may feed other transmissive fractures or will channel subglacial flow to the margin of the ice sheet to be discharged as glaciofluvial outwash. Groundwater flows may be many times that of the present day case.

For the early advance of the ice sheet (3a & b), groundwater chemistry will be similar to that of the periglacial scenario (2) although glacial runoff may locally infiltrate the uppermost parts of the bedrock. As the ice advances and basal meltwater is produced (3c), any meltwater infiltrating the bedrock will tend to dilute the existing groundwaters, particularly any previous deep brines or saline waters concentrated beneath the recently melted permafrost. Dilute, oxidising glacial waters will penetrate to increasing depth as the ice sheet grows and hydraulic pressures increase. Any radionuclides stored beneath the melting permafrost will tend to be flushed out towards the east and will then enter the subglacial discharge to either re-enter the groundwater system or be discharged as glacial outwash SE of Äspö.

5.4 Ice maximum extent: c.60,000 years (Scenario 4)

At the glacial culmination the ice sheet is expected to reach a thickness of 2,200 m at Äspö.

Downwarping due to ice loading may be c.500 m in the Äspö area at this time. Confinement by the large thickness of ice may prevent tectonic stress release during the entire period of glacial loading (e.g. Johnston, 1987). Loading stresses are expected to cause maximum closure of the sub-horizontal Ävrö Zone. The effect of stresses on the near-vertical fracture zones is unclear due to the play-off between horizontal stresses acing to close, and high pore water pressures acting to open, the fractures.

It is considered that Äspö will still lie beneath the melting zone of the ice sheet at this time and hence the bedrock will continue to receive subglacial discharge (e.g. Boulton & Payne, 1992). High infiltration rates are therefore expected to continue into localised transmissive zones. Estimates of mean erosion of only a few tens of metres for past glaciations in Sweden and Finland (e.g. Okko, 1964) suggest that erosion is unlikely to affect significantly the subglacial topography and hence the hydrogeology of the Äspö area, although localised erosive channels may form. Hydraulic connections between the surface of the ice sheet and its base are considered unlikely due to the sealing of crevasses and fissures by ice flow, such that the entire meltwater supply will arise from basal melting of the ice sheet. The transmissivity of the Ävrö Zone is expected to be reduced due to the high vertical stress. Therefore, groundwater flow may be localised above and not within this zone. NE-1 and EW-1 are still considered to be transmissive due to the play-off between increased pore water pressures and moderately high horizontal stresses.

The chemistry of the groundwater will reflect the continued high input of dilute, oxygenating glacial meltwater into the fracture systems.

5.5 Glacial retreat: 60,000 - 70,000 years (Scenarios 5a-d)

The ice sheet is considered to retreat in seawater across Äspö, and hence the ice margin is likely to be in the form of an ice cliff of the order of 100-200m high (e.g. Skinner & Porter, 1987). Permafrost will not become re-established since the area will be beneath the sea. When the ice front has retreated to the site of Stockholm, the sea level is predicted to be 80-100 m above the present day coastline at Äspö. Rapid retreat may be facilitated by the process of frontal calving at the seaward margin. Glacial deposition is considered to be more significant than at other times in the glacial cycle, although it is not considered to modify topography significantly.

The bedrock will begin to uplift and undergo extension, releasing the stresses which had built up during loading by the ice sheet (e.g. Muir Wood, 1993). Certain major fracture zones may be reactivated, although it is not possible to predict which ones are likely to be affected. For the purposes of the scenario exercise it is assumed that NE-1 undergoes reactivation with a normal sense of displacement (5c). This displacement would be consistent with greater uplift to the SE (due to rapid retreat of the ice sheet across the Baltic area) and overall extension which is likely to accompany rebound. Normal fault reactivation is likely to induce coseismic compression of the local bedrock (Muir Wood, 1993). The Ävrö Zone is expected to become more transmissive as the loading stresses decrease.

High rates of basal melting will continue as the warm-based ice sheet retreats across Äspö. Groundwater flow will continue to be directed downwards and outwards (5a,b). The hydraulic gradient across the ice cliff margin will be very high, causing strong upward flow and discharge of groundwater in front of the margin (5c) as the ice cliff passes over Äspö. Possible hydraulic connections between the surface of the ice sheet in relation to the unstable calving of the ice sheet into water or at the margins of localised ice streams may increase the potential subglacial discharge of meltwater by up to two orders of magnitude by introducing surface meltwater to the base of the ice sheet. As the ice sheet retreats further, the sea will cover the Äspö area. The subglacial water input into the groundwater system will be dramatically reduced as the ice front retreats. Reactivation of NE-1 during deglaciation and the associated crustal rebound, may increase its transmissivity. The overall state of extension of the bedrock as it rebounds may tend to increase its porosity and hence induce recharge (5d). However, in the vicinity of NE-1, coseismic compression could induce short-term (days to months) discharge of groundwater (Muir Wood, 1992).

As the ice sheet retreats the marine-fresh water interface is considered to occur in a landward thinning wedge modified by the transmissive fracture system which lags behind the ice margin (5c,d). Once the ice sheet has moved well away from Äspö, sea water may begin to slowly infiltrate the Äspö bedrock. Glacial melt water will contribute less to the groundwater system as the ice sheet retreats away.

5.6 Damming of the Baltic: c.70,000 - 71,000 years (Scenarios 6a-b)

There is a possibility that the Baltic will become dammed during the later stages of deglaciation to 10-30 m above sea level (e.g. Eronen, 1988) (6a). This will cause the waters of the Baltic to become increasingly dilute with time as the meltwater input increases. Incursion of the sea is then predicted (6b), possibly by a jökulhlaup event (Björnsson ,1974), or several events, where water escapes rapidly from the dammed lake after breakdown of the ice barrier or depositional topography causing damming.

Major fracture zones may continue to be reactivated, although no further fault movements are considered in these scenarios. The Ävrö Zone is expected to continue to open-up as rebound continues. A jökulhlaup event may induce seismic activity and reactivation of fracture zones, although such activity is not included in this scenario. The dammed lake level will exert a constant hydraulic head over the region and hence groundwater flow will be low. By comparison with the development of the Baltic Ice Lake during the last glaciation (Eronen, 1988) the path of water escape may be to the north of Äspö such that the Äspö region is not directly affected. However, there is no reason to believe that breaching of the ice dam could not occur in the Äspö vicinity. Damming of the Baltic will cause the Baltic's waters to become increasingly dilute with time as the meltwater input increases. Hence, the low density lake water will not infiltrate to great depths and will tend to form a shallow lens above the more saline groundwater beneath. Re-establishment of the sea across the area would lead to more saline waters gradually infiltrating into the bedrock, encouraged by continued expansion of the bedrock during glacial rebound.

5.7 Interglacial; 71,000 -75,000 years (Scenario 1 or 7)

By c.75,000 years from now, permafrost may only exist in northern Sweden and the climate at Äspö may be similar to that present in northern Scandinavia. Up to 500 m of post-glacial uplift may occur in the Äspö region. This scenario assumes that hydrogeological, hydrochemical and stress conditions fully revert to those of the present-day Base Case before the following glacial period starts. However, at this time it is still expected that sea level will be higher than at the present day by 10-20m, such that conditions at Äspö will not revert fully to those of the present day Base Case before the following glacial period starts. A separate scenario, the Alternative Interglacial Scenario (7) may be used to describe this period. The predicted high sea level during this interglacial implies that flushing of groundwater by meteoric waters and deep saline waters during post-glacial uplift and subaerial exposure will not occur to the extent experienced by the Äspö area prior to the present day. Marine waters may therefore provide a major component of the groundwater beneath Äspö at the start of the next glacial cycle.

6. The Central Climate Change Scenario: Impacts on the Disposal system

The scenario sequence outlined above for the single glacial cycle has been extended and modified in order to represent the glacial evolution of the Äspö area over the next 120,000 years. Estimates of key parameter values and potential relative changes in magnitude at different times in the future, have been compiled for input into performance assessment modelling (Figure 4). Values are approximate and are intended only to give an insight into potential relative temporal changes. Stress data have been taken from modelling undertaken for the SITE 94 project (KTH Group, pers comm., 1994).

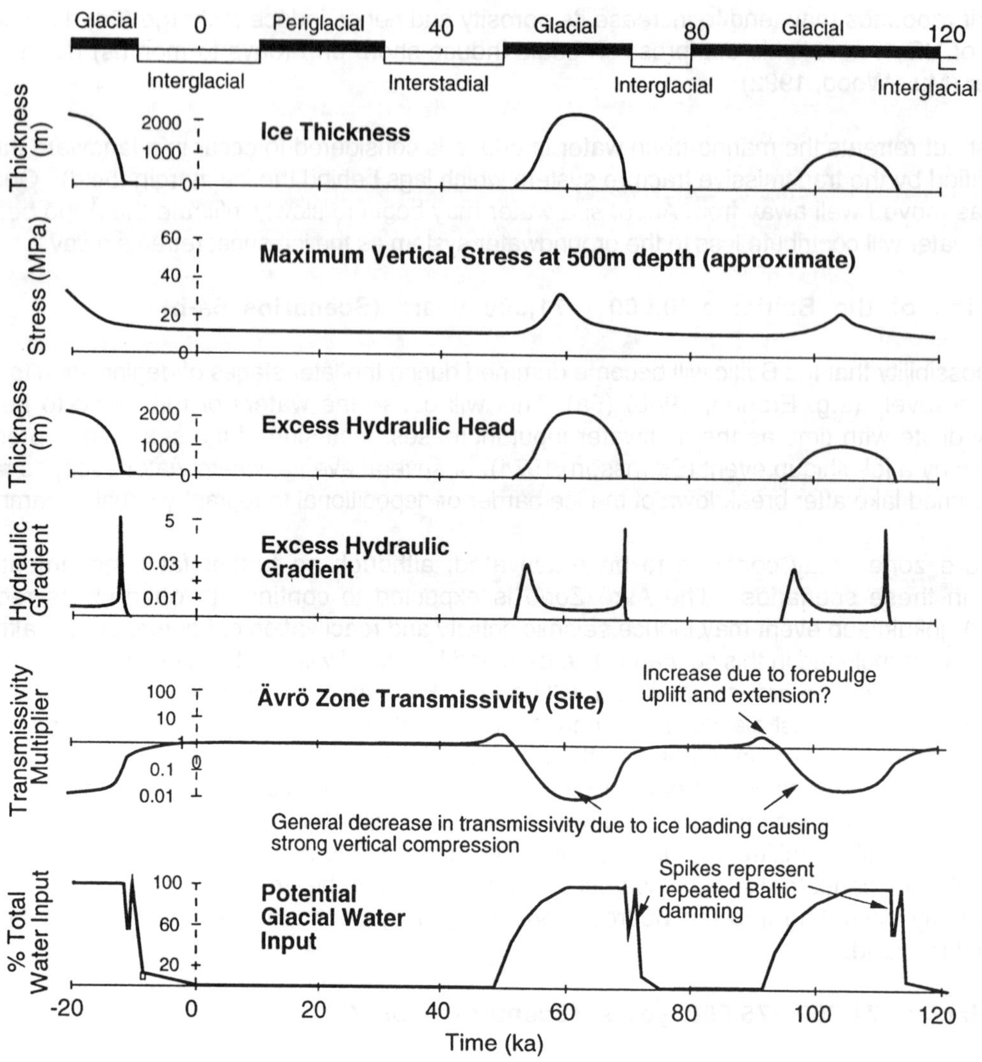

Figure 4. **Graphical representation of potential changes in some climatic, stress, hydraulic and hydrochemical parameters over the next 120,000 years at the Äspö site. Vertical axes are only approximate.**

7. Summary and Conclusions

It has been the intention that the Central Scenario illustrates potentially significant aspects of climate change for analysis of the future safety of a repository for spent fuel. A number of glacial cycles are considered to affect the Äspö region during the next 120,000 years, during which the mechanical and hydrogeological stability of a potential repository would be affected by such factors as ice loading, permafrost development, temperature changes, biosphere changes and sea level changes. The main periods of groundwater discharge, enhanced groundwater flow and hydrochemical change, which are of considerable significance to the performance assessment of a potential repository, are considered to occur during the early advance and late retreat phases of a glaciation. The latter discharge phase is considered to be more significant than the former phase which is inhibited by the development of permafrost in the proglacial zone. During ice sheet development oxygenated waters are forced into the bedrock, particularly via more transmissive fracture zones, by a combination of head increases and permafrost sealing. Flushing by dilute meltwaters continues until the ice sheet retreats from the Äspö region. During ice loading the bedrock undergoes bulk compression whereas subsequent glacial unloading causes expansion of the rock volume.

The numerous estimates and assumptions made during the development of the Äspö scenarios have been outlined. As a result of this study, it has been found that there appears to be a general lack of observational data on the direct effects of an ice sheet on the hydrogeology of the underlying bedrock. Although this is understandable on practical grounds, it means that there is a dependence on glaciological modelling. Glacial models for past and future glaciations that have been identified during this study (e.g. Boulton & Payne, 1992) appear to deal with sedimentary formations comprising the subglacial bedrock. The reaction of fractured crystalline bedrock may be significantly different to that of interbedded aquifers and aquitards. Subglacial meltwater tunnels are more likely to develop above crystalline bedrock where the glacial discharge cannot be accommodated fully by groundwater flow.

References

Ahlbom, K, Äikäs, T & Ericsson, L O, 1991. SKB/TVO Ice Age Scenario. SKB Technical Report, 91-32.

Anundsen, K, 1993. Crustal Movements and the Late Weichselian ice sheet in Norway. SKI Technical Report KAN 3 (93) 12.

Berger, A., Gallée, H., Fichefet, T., and Tricot, C. 1989. Testing the astronomical theory with a coupled climate-ice sheet model. In : Global and Planetary Change. Ed. Labeyrie, L.

Björck, S & Svensson, N-O, 1992. Climatic change and uplift patterns - past, present and future. SKB Technical Report, 92-38.

Björnsson, H, 1974. Explanation of jökulhlaups from Grimsvotn, Vatnajökull, Iceland. Jökull, 24, 1-26.

Böse, M, 1990. Reconstruction of ice flow directions south of the Baltic Sea during the Weichselian glaciations. Boreas, 19, 217-226.

Boulton, G S, 1991. Proposed approach to time-dependent or “event-scenario” modelling of future glaciation in Sweden. SKB Arbetstrapport, 91-27.

Boulton, G S & Jones, A S, 1979. Stability of temperate ice sheets resting on beds of deformable sediment. Journal of Glaciology, 24, 29-43.

Boulton, G S & Payne, A, 1992. Simulation of the European ice sheet through the last glacial cycle and prediction of future glaciation. SKB Technical Report, 93-14.

Chapman, N A, Andersson, J, Skagius, K, Wene, C-O, Wiborgh, M & Wingefors, S, in prep. Devising Scenarios for future repository evloution: A rigorous methodology. Materials Research Society Publication: Proceedings of the Scientific Basis for Nuclear Waste Management Kyoto Meeting, 1994.

Clark, P U & Walder, J S, 1994. Subglacial drainage, eskers, and deforming beds beneath the Laurentide and Eurasian ice sheets. Geological Society of America Bulletin, 106, 304-314.

Drewry, D, 1986. Glacial Geologic Processes. Edward Arnold. 276 pp.
Embleton, C & King, C A M, 1975. Glacial Geomorphology, Arnold Publishers Ltd.

Eronen, M, 1988. A scrutiny of the late Quaternary history of the Baltic Sea. Geological Survey of Finland, Special Paper, 6, 11-18.

Fairbanks R G 1989. A 17,000-year glacio-eustatic sea level record: influence of glacial melting rates on the Younger Dryas event and deep-ocean circulation. Nature 342, pp. 637-642.

Fjeldskaar, W., and Cathles, L. 1991. Rheology of mantle and lithosphere inferred from postglacial uplift in Fennoscandia. In Glacial isostasy, sea-level and mantle rheology. Sabadini, R., Lambeck, K. and Boschi, E. (Eds).Nato ASI Series C. vol 334. pp 1-19.

Hindmarsh, R C A, Boulton, G S & Hutter, K, 1989. Modes of operation of thermomechanically coupled ice sheets. Annals of Glaciology, 12, 57-69.

Imbrie, J & Imbrie, J Z, 1980. Modelling the climatic response to orbital variations. Science, 207, 943-953.

Johnston, A, 1987. Supression of earthquakes by large continental ice sheets. Nature, 330, 467-69.

King, L M, Chapman, N A, Kautsky, F, de Marsily, G & Svensson, N-O, 1994 (in prep). The Central Scenario for SITE 94. SKI Report, SKI 94:16.

McEwen, T J & de Marsily, G, 1991. The Potential Significance of Permafrost to the Behaviour of a Deep Radiocative Waste Repository. SKI Technical Report, 91:8.

Mörner, N A, 1989. The Fennoscandian uplift and Late Cenozoic geodynamics: geological evidence. Geojournal, 3.3, 287-318.

Muir Wood, R, 1992. Earthquakes, water and underground waste disposal. In: Waste Disposal and Geology; Scientific Perspectives. Proceedings of Workshop WC-1 of the 29th International Geological Congress, Tokyo, 169-192.

Muir Wood, R, 1993. A review of the seismotectonics of Sweden. SKB Technical Report, 93-13.

Okko, V, 1964. Maaperä. In: Rankama, K (ed.). Suomen geologia. Kirjayhtymä, Helsinki.

Shen, B & Stephansson, O, 1990. 3DEC Mechanical and Thermo-Mechanical Analyisis of Glaciation and Thermal Loading of a Waste Repository. SKI Technical Report, 90:3.

Skinner, B J & Porter, S C, 1987. Physical geology. 750 pp. John Wiley & Sons.

Smellie, J & Laaksoharju, M, 1992. The Äspö Hard Rock Laboratory: Final evaluation of the hydrogeochemical pre-investigations in relation to existing geologic and hydraulic conditions. SKB Technical Report, 92-31.

Svensson N-O 1989. Late Weichselian and early Holocene shore displacement in the central Baltic, based on dtratigraphical and morphological records from eastern Småland and Gotland, Sweden. Lundqua Thesis 25.

Svensson N-O 1991. Late Weichselian and early Holocene shore displacement in the central Baltic sea. Quaternary International 9, pp. 7-26.

Walcott, R I, 1970. Isostatic response to loading of the crust in Canada. Canadian Journal of Earth Sciences, 7, 716-726.

Research by Nirex into Long-Term Geological Changes at Sellafield

R. Chaplow and A.J. Hooper
United Kingdom Nirex Ltd.
(United Kingdom)

Abstract

Nirex is investigating an area close to Sellafield in Cumbria, England as a potential site for a deep repository for intermediate- and some low-level radioactive wastes. Studies into potential long-term geological change have begun to identify, in particular , evidence of past changes to the groundwater flow system as a basis for estimating possible future natural changes to the system. Studies of the past history of subsidence and uplift, the evolution of the fractures controlling groundwater flow, the Quaternary evolution of the area and palaeohydrogeological studies are providing an understanding of the way the site has evolved, particularly during the last 60 million years since the end of the Cretaceous period. Although the studies are at an early stage, and much work remains to be done, encouraging progress is being made and the site is being shown to possess a fair degree of stability in terms of the groundwater flow system

Introduction

Nirex is responsible for providing and managing a national disposal facility for solid intermediate-level (ILW) and low-level radioactive waste. UK government policy is to dispose of these wastes in a deep underground repository.

This paper describes some aspects of the research being undertaken by Nirex into long-term geological changes at the Sellafield site and is sub-divided into the following parts:

1. an introductory section which describes the approach being taken by Nirex to radioactive waste disposal. The location of the site at Sellafield, the disposal concept being adopted and the pathways for radionuclide transport are described;

2. the geological and hydrogeological characteristics of the site are described; and

3. some of the approaches being adopted towards studying geological change are described.

Nirex's Approach to Radioactive Waste Disposal

Sellafield is located near the western coastline of Cumbria, England, some 500 km north-west of London. Nirex, in common with disposal organisations in other countries, has developed a concept of deep geological disposal of radioactive waste which uses a multi-barrier containment system . It is envisaged that caverns will be excavated at depth in a stable geological environment. Wastes, set in steel or concrete packages, will be placed in the caverns which will then be backfilled with a cement-based material.

The concept makes use of both engineered and natural barriers, working in conjunction, to achieve the necessary degree of long-term waste isolation and containment. It includes a simple and robust engineered system situated at adequate depth in a region of naturally low groundwater flow.

A physical barrier within the repository will be provided by the steel and concrete packaging within which the radioactive material is immobilised. Experimental results suggest that the packaging should itself provide a high degree of containment over several hundred years. However, the long-term containment properties of the engineered system in respect of radionuclides dissolved in groundwater will stem from the establishment of uniform chemical conditions and high sorption capacity across the repository. This will be achieved by surrounding waste packages with the required amount of a cement-based backfill. This backfill has been carefully specified to fulfil a number of requirements, namely :

1. long-term maintenance of alkaline porewater chemistry in order to suppress the solubility of key radionuclides under the prevailing conditions of groundwater flow and geochemistry;

2. long-term maintenance of a high active surface area for sorption of key radionuclides; and

3. relatively high hydraulic conductivity and porosity to ensure homogeneous performance, in that localised concentrations of materials in wastes would not

exhaust the desired chemical conditioning and thereby locally reduce the containment performance.

To date, development of Nirex assessment models has focused on three major pathways:

1. transport of radionuclides in groundwater;

2. migration of radionuclides in gases; and

3. return of radionuclides to the environment as a result of natural disruptive events or inadvertent human intrusion.

Assessment studies have indicated that, of these, the groundwater pathway is the most important in defining the performance of the containment system. Long-term geological changes can potentially impact the effectiveness of geological containment by changing the groundwater flow system.

UK Government policy on radioactive waste management does not envisage any time cut-off for assessments of safety of disposal, but recognises the questionable nature of quantitative assessments beyond times of at most a few million years.

Geological and Hydrogeological Characteristics of the Sellafield Site

Sellafield is located on the eastern margins of the East Irish Sea Basin which occurs between the Lake District Massif to the east and the Isle of Man to the west. The proximity of this basin has had, and continues to have, a major influence on the characteristics of the Sellafield site and the way it is evolving.

The patterns of subsidence and uplift associated with basin evolution, the successive phases of faulting, mineralisation and diagenesis, and the development and subsequent movements of brines exert major influences on the long-term evolution of the site.

Subsequent to the main periods of basin subsidence and uplift, the evolution of the site has been strongly affected by glaciation. The upland areas of the Lake District Massif have been subjected to extensive erosion with the development of characteristic glacial landforms. In the lower lying areas around the central uplands, a discontinuous veneer of Quaternary sediments, up to 180 metres thick, mask the solid rock. A basal till was formed during the main late Devensian Glaciation which reached its maximum extent over Britain some 18,000 years ago. Glaciofluvial sands and gravels and glacilacustrine clays and silts were deposited by outwash meltwaters during retreat of the main ice sheet, probably about 14,000 years ago. An onshore ice readvance known as the "Scottish Readvance" resulted in deformation of pre-existing sediments near to the present coastline.

Glaciation had a profound effect on the Sellafield area. There was probably some 2 km of ice over the site, and the sea level fell to the extent that the coastline of the Irish Sea retreated to the west of the Isle of Man. Sellafield thus became an inland, rather than a coastal site.

There is strong evidence that Britain is likely to experience further glaciations during the time periods covered by the post-closure performance assessments for a potential repository at Sellafield. Hence, the impacts on the site of glacial advance and retreat is an important consideration in the studies of long-term change.

At the Sellafield site the Ordovician volcanic rocks of the Borrowdale Volcanic Group (BVG) form the uplands of the Lake District massif and the basement of the younger sedimentary sequence deposited in the East Irish Sea Basin, at least in the area close to Sellafield. The potential repository would be located in the BVG at a depth of around 750 to 1,000 metres, although the final depth is still to be confirmed if the site is shown to be suitable.

The influence of the East Irish Sea Basin on the sedimentary sequence close to Sellafield is shown by the way in which this sequence, comprising Carboniferous Limestone, a breccia (the Brockram), evaporites and shales of Permian age and the overlying Triassic Sherwood Sandstone Group, is truncated and the sequence of deposition has been controlled by the pattern of faulting associated with the Fleming Hall Fault Zone, located to the west of the Potential Repository Zone (PRZ).

Approximately 2 km offshore, the Triassic sandstones are overlain by the Mercia Mudstone. The importance of this formation is that it contains extensive deposits of halite which are believed to be the source of the brines which are encountered at depth in the eastern part of the Sellafield site.

Key Issues Affecting Long-Term Change

In addressing the long-term geological changes at the site there are two key issues:

1. the nature of the main processes which are likely to affect the site in the future and affect the groundwater flow system; and

2. the evidence which can be obtained for the likely impact of these processes on groundwater flow and radionuclide transport.

A geological principle which guides much of the thinking on geological change is that of uniformitarianism: that the present can be used to interpret the past. This principle leads directly to the concept that processes operating at present and in the past will operate in the future, and produce similar results over the timescales of interest to repository post-closure performance assessment studies.

Current Nirex Studies of Geological Change

A wide range of geological, hydrogeological and geochemical studies are being undertaken to develop an understanding of the long-term evolution of the site as a guide to potential future changes. The major components of these studies are:

1. geological studies to determine the history of the development of the Sellafield area, including history of deposition, burial and uplift;

2. studies to elucidate the history of the development of the fractures (faults, joints, etc.);

3. studies of the Quaternary deposits, particularly with regard to the information they can provide on the history of glaciation and deglaciation, sea level changes and the possible occurrence of recent fault movements; and

4. palaeohydrogeological studies to establish the way in which the groundwater system has evolved through time.

Tectonic and climatic processes are likely to affect any site in the UK over the time periods we need to consider. The issue is therefore one of establishing a rational basis for estimating the impact of these changes so that they can be given appropriate recognition in the assessment of the post-closure performance of a repository.

Each of these aspects of geological change are briefly discussed in the succeeding sections of this paper.

History of Development of the Sellafield Area

The geological studies undertaken by Nirex have enabled the reconstruction of the history of deposition, uplift and erosion which have occurred since the initiation of the East Irish Sea Basin, even though rocks younger than the Jurassic, other than the Quaternary deposits, have been removed by erosion.
Basin subsidence was initiated in the Carboniferous and was particularly well marked through the Permian and into the Triassic. Uplift was initiated at the end of the Cretaceous, some 60 million years ago, and this uplift and erosion may well be continuing. Thus the rocks at Sellafield appear to have been subjected to a progressively reducing overburden stress for the last 60 million years. Estimates of uplift, derived from apatite fission track analysis and other studies, suggest that the Sellafield area may have experienced something in the order of 2,000 metres of uplift since the end of the Cretaceous.

Fracture Studies

A second aspect of the studies of the geological history has related to dating the periods of faulting that have affected the area. Traditionally, fault movements were dated indirectly by determining the ages of rocks offset by the faults and the ages of undisturbed, cross-cutting (e.g. dykes or mineral veins) or overlying strata. In recent years, a growing number of techniques have been applied to enable direct dating of fault rocks using a similar approach to that adopted for dating the fracture mineralisation.

Using these new techniques, two groups of fault rocks have been dated, namely:

1. three samples from a fault encountered in Boreholes 2 and 4; and

2. three samples from Borehole 3 and 4 which shared many features but were markedly different from the first group.

The samples from the first group gave ages for the fault wall rock ranging from 255 to 315 million years. Fault gouge samples were dated at between 118 and 146 million years and were considered to represent minimum ages for fault activation and reactivation along different planes in the same fault zone.

An authigenic illite clay in the second group gave a minimum age of 212 million years, possibly representing a period of enhanced tectonic activity during the opening of the East Irish Sea Basin in the Triassic. The minimum age of authigenic illite-smectite clay also in the second group was estimated to be 60 million years. This is not thought to be related to a faulting episode, but may be due to a thermal event at that time in the Tertiary.

Hydrogeological testing carried out in the Boreholes has identified those sections of the borehole wall through which groundwater can be induced to flow. The observation was made that groundwater flow in the BVG, and, to some extent in other formations, occurred through a small subset of fractures. It was thus shown that the dominant influence on groundwater flow through the BVG was not the complete set of fractures but rather this small subset of fractures. The zones in the boreholes through which groundwater flow can be induced are called 'flow zones'. Using data from the boreholes and from analysis of the rock cores, these flow zones have been characterised. In some cases the flow is found to occur along one or more fractures in the rock. These fractures are termed 'flowing fractures'. In other cases, particularly in the Brockram and sandstones, flow cannot be assigned to particular fractures since it may, for example, be occurring partially or completely through the rock matrix. In these cases the flow zone is referred to as a 'flowing feature'.

Having shown the importance of the flow zones in controlling the hydraulic conductivity of the BVG, the next important issue is to characterise them so that they can be appropriately incorporated into groundwater flow models of the site.

The studies so far undertaken by Nirex have characterised the flow zones in nine boreholes within the PRZ, namely Boreholes 2, 4, 5, RCF1, RCF2, RCF3, RCM1, RCM2 and RCM3 .

In total, only 154 flow zones have been identified in the boreholes which together amounted to over 8,000 metres of drilling. This is about 0.5% of the total number of fractures encountered in the boreholes. Fifty percent of the flow zones occur in the sandstone, 8% in the Brockram and the remaining 42% in the BVG. Thus, only 64 flow zones have been identified in the BVG in the area studied.

The analysis of these flow zones has been carried out to identify specific features responsible for the inflow of water and to investigate the mineralogical and petrological character of these and adjacent features seen in the borehole cores. In particular, the mineralisation, distribution, orientation and character of the features have been examined.

The minerals and fluid inclusions which occur in the fractures provide information which can be used to interpret the time when the fractures were created and their history in terms of further periods of movement which may have occurred since their formation. Stable isotope signatures and fluid inclusion characteristics of the fracture minerals provide an insight into the chemistry of the mineralising fluids and the conditions under which mineralisation occurred. The investigations of fracture mineralisation commenced with a general review of the history of mineralisation and the establishment of a series of mineralisation episodes which have affected the rocks at different times in their history and have resulted in the deposition of different minerals within the fractures. Subsequently, the studies of fracture mineralisation have concentrated on the intervals containing flow zones.

These studies have enabled the establishment of a well-defined history of fracture movement and discrete mineralisation events affecting the site. Nine broad but discrete mineralisation episodes (designated ME1 to ME9) have been recognised. These can be correlated with regional patterns of mineralisation and diagenesis, both in the East Irish Sea Basin and in the Lake District. Limited potassium-argon dating of illite clay mineralisation in fault gouge, and correlations with regional patterns of mineralisation have enabled the periods of fracture development and mineralisation to be dated.

In most cases there appears to be no particular relationship between fracture orientation, fracture type and mineralisation episode. Each mineralisation episode, with the exception of ME2, has reactivated and reutilised the pre-existing fractures and veins of earlier episodes. This is because the carbonate- or hematite-dominated fracture-fill is relatively weakly bonded to adjacent wallrock and therefore the veins represent planes of weakness, which are readily reactivated in preference to the formation of new fractures.

The following conclusions have been reached from these studies of the flow zones :

1. flow zones are not associated with a single, consistent geological characteristic in the boreholes studied;

2. the flow zones in the Sherwood Sandstone Group are largely matrix flow with some contribution from fractures. For those flow zones associated with fractures, there is a tendency towards shallow south-west dipping orientations, which reflects the dip of the bedding in the rocks;

3. in the Brockram the flow zones are associated with either matrix flow or fracture flow;

4. in the BVG the flow zones are mostly associated with mineralised flowing fractures. When flow in the BVG can be associated with specific flowing fractures, there is a slight tendency for these to be oriented north-east dipping;

5. there is a broad relationship between flow zones and the occurrence of ME6 carbonate mineralisation. Of the 78 flowing fractures identified and examined so far, more than 80% were associated with carbonate/hematite mineralisation. Seventy two have been assigned to a particular mineralisation episode and, of these, 48 (67%) were assigned to mineralisation episode ME6;

6. there is no obvious relationship between flow zones and faulting in the St. Bees Sandstone. Within the BVG, 20% of the flow zones occur within fault rock and 47% within 5 metres of a fault. The association between ME6 mineralisation and faulting is variable between individual boreholes;

7. there is a strong relationship between flow zones and the presence of ME9 late calcite mineralisation. ME9 post-dates all observed fault structures observed in borehole cores. The type of late calcite crystals found appears to be related to the current salinity of the groundwater. If the late calcite can be successfully dated, it may help to indicate the period of time for which the present salinity profile has existed;

8. the close association between flow zones, proximity to faults, ME6 mineralisation and north-east dipping fractures suggests the possibility that groundwater flow in the BVG is related more to the larger scale distribution of ME6 mineralisation in the rock mass, rather than the type, orientation or intensity of individual flowing features or fractures; and

9. there is an apparent linear reduction in measured hydraulic conductivity values in the flowing fractures with depth.

Relating these various conclusions to the understanding of the geological history of the site suggests that the majority of the flowing fractures probably owe their origin to fracture movements which occurred around the early to middle Triassic, some 200 million years ago when the ME6 mineralisation episode occurred. These fractures probably remained as flowing features during subsequent periods of fault movement and mineralisation. The last significant period of fault movement occurred around 100 million years ago as indicated by the dating of clays in fault gouges. The flowing fractures appear to have remained hydraulically active subsequently as indicated by the development of ME9 late calcite. Such mineral precipitation, and associated dissolution, is likely to have altered the specific characteristics of individual fractures and may have caused some adjustments to the networks of connected fractures over time. However, the evidence so far supports the view that the ME6 fracture network has provided the framework for the network of connected fractures and that this framework has persisted for some 200 million years.

Quaternary Studies

During the latter part of the Quaternary, the British Isles have been glaciated on several occasions. About ten more glaciations of comparable extent are anticipated in the next million years, subject to the influence of greenhouse gas warming on long-term climate change.

The results of the Quaternary Characterisation studies available so far have indicated that:

1. the area has been glaciated and subjected to changes in sea level over the last tens of thousands of years; and

2. the Quaternary sediments show evidence of dislocations associated with ice movement (glacitectonic structures). However, despite extensive searching using remote sensing and high resolution seismic surveys, no evidence has been found of tectonic faults having involved both the bedrock and the Quaternary deposits themselves.

Palaeohydrogeological Studies

The geochemical characteristics of the water provide information on how the groundwater system has evolved over time, including providing evidence of past groundwater flow directions and mixing processes which are occurring within the system. The interpretation of groundwater flow and mixing processes that arise from the geochemical studies thus provide independent evidence that can be compared with the predictions of groundwater flow derived from consideration of the physical characteristics of hydraulic conductivity and head.

Nirex has identified three groundwater regimes. These are:

1. a shallow, fresh-water occurring throughout the area within the 'Coastal Plain Regime';

2. a saline water with a high bromine to chlorine ratio (relative to the brines in iii. below) occurring at depth towards the east of the area, within the 'Hills and Basement Regime'; and

3. a deep brine with a lower bromine to chlorine ratio occurring at depth in the west of the area, within the 'East Irish Sea Regime'.

The recent data have indicated that the freshwater can be subdivided into two types distinguished on the basis of the stable oxygen isotope data. This newly identified type occurs locally beneath the Coastal Plain Regime where the freshwater, on the basis of the measured stable oxygen and hydrogen isotopes, is different from the shallow freshwater. It is believed that this deeper water was recharged under colder climate conditions than currently exists, possibly during the Pleistocene.

The recognition of this fourth component, and the way in which it has enabled all the groundwater compositions so far observed at the site to be attributed to mixing between the four identified components, represents a significant advance in the understanding of the geochemistry of the Site. This is because it provides confidence that the pattern of groundwater type distribution has been revealed and that there are no longer any observed compositions that cannot be explained in terms of well defined mixing relationships.

The saline transition zone, which marks the interface between the fresh water above and the saline water below, is relatively sharp in the PRZ and generally occurs in the Brockram and basal sandstones. To the south and west of the PRZ the saline transition zone is more diffuse.

A good indicator of the rate of groundwater flow through the system is provided by the length of time that the water has been resident in the ground. Information can also be obtained for the residence time of the solutes dissolved in the water. The best evidence for residence times for water and solutes in the BVG in the PRZ comes from stable isotopes and chlorine-36 data respectively. The stable isotopes (oxygen and hydrogen) together with data from analysis of the noble gases in the water, indicate the water has been recharged (entered the ground) under colder climate conditions, during the Pleistocene (up to 1.6 million years ago). The chlorine-36 data indicate that the chloride in the water in the BVG in the PRZ has had a long residence time, possibly over 1.5 million years. Preliminary helium data also indicate long residence times of similar magnitudes.

The latest interpretation of the geochemical data is thus indicating:

1. a consistent pattern for the distribution of groundwater types;

2. all the observed groundwater compositions can be attributed to mixing between four identified groundwater components; and

3. long residence times for the water and solutes, possibly up to 1.5 million years. This is shown by four independent sources of data: stable oxygen/hydrogen isotopes, noble gases, chlorine-36 and helium data. Evidence of mixing between components, coupled with long residence times suggest low flow of groundwater through the BVG at depth and a high potential for dilution and dispersion occurring in the ground of any residual mobile or long-lived radionuclides from a repository in groundwater that may ultimately reach the surface environment.

The palaeohydrogeological aspects of the geochemical studies have indicated that the deeper freshwater component was possibly recharged in a cold (glacial or periglacial) phase of the Pleistocene, with the shallower component being recharged since that time. However, no evidence has been found to indicate that there has been extensive flushing of the water from depth during the last glaciation (10,000 to 26,000 years ago) since the chlorine-36 and helium data have indicated residence times for solutes in the order of 1.5 million years.

Conclusions

The studies undertaken by Nirex are focused on understanding natural processes and the ways in which they have influenced the evolution of the site. Particular attention has been directed towards those processes which may influence groundwater flow within the geosphere.

Efforts are being made to develop models of the site evolution which can be tested by seeking to predict current observations on the basis of processes which have influenced the site in the past.

The studies of geological change are at an early stage. The results so far provide encouragement in the progress being made. However, much work is still required to develop an adequate understanding of long-term geological change.

Studies of the Geologic Stability of the Canadian Shield For Siting a Nuclear Fuel Waste Disposal Facility

C.C. Davison, A. Brown, R. Everitt and M. Gascoyne
AECL Research
Whiteshell Laboratories
Pinawa, Manitoba
Canada
R0E 1L0

Abstract

AECL Research has undertaken a variety of studies to address issues related to understanding and assessing the geologic stability of plutonic rocks of the Canadian Shield to assist in siting a nuclear fuel waste disposal facility. These studies include: monitoring the current levels of seismicity of the Shield and using this information to bound the probability of the site specific occurrence of an earthquake large enough to affect the integrity of the geosphere surrounding a disposal vault; and, examinations of the age, origin and evolutionary history of fractures found in plutonic rocks of the Shield. A case study is presented for a well-characterized fracture at the site of AECL's Underground Research Laboratory (URL) to illustrate an approach for examining the history of fracture reactivation and propogation at a specific site.

1. Introduction

The Canadian Nuclear Fuel Waste Management Program (CNFWMP) has been assessing the concept of disposing of nuclear fuel waste in a vault 500m to 1000m deep within plutonic rock of a stable portion of the Canadian Shield. Although parts of the Canadian Shield are currently among the most seismically stable regions of the world, infrequent earthquakes have been recorded in these stable areas. Therefore, investigations for siting a disposal vault in the Canadian Shield need to include an assessment of the geologic stability of the rocks at otherwise potentially suitable sites. This assessment must consider the current and past levels of seismic stability, estimate the possible future stability, and evaluate the significance of future earthquakes on the integrity of the disposal vault (Martin et al, 1994) and surrounding geosphere (Brown et al, 1994). In Canada, the regulatory requirement is a quantitative risk assessment for 10,000 years following vault closure and a qualitative assessment beyond 10,000 years.

2. Seismic Stability of the Canadian Shield

The Archean portion of the Canadian Shield is the least seismically active portion of the North American plate and one of the least seismically active regions in the world. However, despite this very low seismic activity, earthquakes do occur from time to time and it is important to identify zones of differing seismicity within the Shield to assist in siting a disposal vault for nuclear fuel wastes. Since the beginning of the CNFWMP in 1978, AECL has maintained a network of seismic monitoring stations in the Ontario portion of the Shield to provide a record of low levels of seismicity that occur. Historically, earthquakes on the Shield are often clustered along geologic structural features - the Kapuskasing structural zone (a deep crustal thrust), the Timiskaming rift, and the Ottawa/St. Lawrence rift system or other ancient rifts or breaks in the North American plate, North Central Quebec and the Ungava area of Northern Quebec. (Figure 1) The region of the Shield including these features has been categorized as having moderate seismic activity (Basham and Cajka 1985). Adams and Basham (1991) concluded that most of the large infrequent earthquakes in the Shield of Canada could be related to the location of ancient rifts (old breaks in the North American plate) formed during previous separations of larger plates or with the current continental margin (one side of the rifts associated with opening the Atlantic Ocean).

In order to evaluate the risk of damage to a disposal vault associated with proximity to active faults, Atkinson and McGuire (1993) have used two related approaches to estimate the annual probability of new fractures caused by earthquakes on an active fault near a disposal vault.

- One approach considered the potential for secondary fracturing to extend to the edge of a disposal vault from an active fault. For this approach Atkinson and McGuire (1993) used a magnitude/frequency relationship for a portion of the Shield given by Atkinson (1992), a magnitude/rupture area relationship based on Wyss (1979), and a magnitude/secondary fracturing relationship based on Bonilla (1970).

- The other approach considered the potential for damage using three different peak ground accelerations (0.2 g, 0.3 g and 0.5 g) as criteria for occurrence of damage. For open excavation at depth, Dowding and Rozen (1978) reported that no damage was observed for peak ground accelerations of less than 0.2 g and only minor damage at peak ground acceleration of less than 0.5 g. Following closure of the disposal vault, even a 0.5 acceleration should overestimate the likelihood of damage by a wide margin because there would no longer be any open excavations.

So long as no active fault was located within 50 m of a disposal vault at a site on the Shield, Atkinson and McGuire (1993) estimated the annual probability of fracturing reaching the disposal vault to be 5×10^{-7} for the most pessimistic assumption for the extent of secondary fracturing (the upperbound of values reported by Bonilla 1970) and the most pessimistic peak ground acceleration damage criteria (0.2 g). For a less pessimistic assumption for secondary fracturing (average values report by Bonilla 1970), the estimated probability was 1 x 10^{-7}. If the geometry of the active faults was known for the latter case, the estimated probability of fracturing reaching the disposal vault could be effectively reduced to zero by keeping the disposal vault more than 1 km away from any active fault that was greater than 2 km in length, and more than 200 m away from any active fault that was greater than 0.5 km in length (Atkinson and McGuire 1993).

3. Methods of Determining the Age, Origin and History of Fracturing at a Site

Pre-existing fault and fracture zones in the rock are weaker than regions of either moderately fractured or sparsely fractured rock. Individually, they are larger than individual fractures in moderately fractured rock. Therefore, we expect any potential future movement associated with earthquakes at or near a disposal site in the Shield would take place on pre-existing faults or fracture zones in the rock and would not create new faults or fracture zones.

In addition to monitoring the current levels of seismicity, the potential for future fracture propagation at a potential disposal site may be assessed by developing an understanding of the fracture history of the rockmass at the site. This history would include the initial fracture formation as well as subsequent episodes of reactivation and propagation of the fractures. These events leave behind a record in the form of overlapping fracture mineral infillings and different alteration assemblages adjacent to the fractures. Therefore, the propagation rate through geological time for a given fracture (or for a fracture network) can be defined by studying the mineral infillings and alteration assemblages of the fractures using the following steps:

- identifying and dating the infilling and wall rock alteration assemblages encountered,
- delineating the extent of these assemblages within a given fracture, and then
- constructing a plot of mineral extent versus mineral age (essentially a simple age map of the fracture surface).

The resulting propagation curve, and the tectonic history under which it developed, may then be used to estimate (at least in a qualitative way):

- the potential for continued fracture development under the present conditions, and
- the tectonic conditions required to accelerate this process (based on past history).

The absolute or relative age of successive (overlapping) mineral assemblages may theoretically be derived from field relationships, microstructural examination, mineral stability fields, isotopic methods and palaeomagnetism.

The principal constraint on the application of this approach at a particular nuclear fuel waste disposal site is the quantity and variety of the fracture infillings and altered wall rock available for analysis. For instance, it is known that only a limited number of fracture infilling types occur in the rock at AECL's Underground Research Laboratory (URL) in comparison to other sites in the Canadian Shield, and that these infillings are generally very thin and discontinuous. Another constraint is the ability to delineate the extent of the infilling and wall rock alteration assemblages along a given fracture. This requires a continuous exposure of the fracture or multiple sample intersections so that the successive mineralogical stages in the development of a fracture can be sampled.

The Room 209 subvertical fracture at the 240 m level of the URL is used here to illustrate the method for estimating a fracture propagation rate curve from examining the mineralogical history of the fracture refillings. This is the best studied of the subvertical fractures at the URL. This fracture has been sampled near its tip by multiple borehole intersections and by tunnel mapping (Lang et al 1988), so that a complete gradation of infillings and alterations is observed (Figure 2). The oldest infilling, chlorite, is overlain by hematite, then carbonate, and finally by clay. The wall rock alteration ranges from strongly hematized (brick red) through pink, and then to unaltered at the fracture tip. The spatial extent of these mineral assemblages is shown in Figure 2. Absolute ages for the various mineral assemblages occurring in the fracture networks in the rock mass at the Underground Research Laboratory have been determined by Kamineni and others (listed in Stone et al 1989, and Syme et al 1993) and are used here in the absence of infilling age data specific to this fracture. A summary of the available fracture infilling age data is presented in Table 1.

4. Case Study of the Room 209 Fracture History

The historical rate of propagation for this vertical fracture can theoretically be estimated by compiling a plot of the mineral assemblage ages versus their respective spatial extent on the fracture. Figure 3 is a simplified cross section of the subvertical Room 209 fracture. The assemblage of minerals occurring as infillings are summarized to the left of the fracture trace while their extent within the fracture is shown by the different line widths. The accompanying wall rock alteration is listed on the right of the fracture trace.

These different fracture infillings and alteration assemblages represent stages in the growth of the fracture. In this case the stages include:

1. fracture formation to point "1", deposition of chlorite within the fracture
2. reactivation and propagation to point "2", deposition of hematite within the existing fracture and in its newly propagated tip.
3. fracture reactivation and propagation to point "3", deposition of carbonate within the existing fracture and in its newly propagated tip.
4. fracture reactivation and propagation to point "4", no visible infilling.

The fracture propagation curve defined by these stages is shown in Figure 4 (a plot of assemblage age versus their relative extent on the fracture surface.

Chlorite, formed prior to 2.1 Ga but after the low-dipping faults (based on crosscutting relationships), is preserved on about 20% of the total fracture surface. It may have once been more continuous as indicated by isolated traces elsewhere on the fracture surface (e.g. borehole 7 shown in Figure 2). Hematite (the high-temperature hydrothermal-type described in Brown et al (1989) formed prior to 2.0 Ga (est.)) overlies the chlorite but also extends beyond it, for a total of about 90% of the fracture surface. The remainder of the fracture contains only minor quantities of low temperature infillings and alteration, whose estimated age is poorly defined but is believed to range approximately as shown.

The granite itself and the various dykes it hosts crystallized and cooled in the interval between 2.7 Ga and 2.3 Ga, while the low-dipping fault zones may have been initiated as early as 2.298 Ga (Brown et al 1989, Stone et al 1989).

From inspection of Figure 4, it is apparent that most of the present-day fracture surface was formed during or before chlorite deposition (about 2.1 Ga ago), and that the rate of subsequent reactivation or propagation of the fracture appears to have rapidly decreased with time.

This preliminary analysis of the history of the subvertical fracture in Room 209 of the Underground Research Laboratory suggests that the large subvertical fractures of the "Room-209 type" in the rock mass at the URL site developed rapidly at the close of the cooling history of the granite, during a period of rapid uplift and erosion. The fractures do not appear to have propagated much during the numerous episodes of deformation, rifting, uplift, deposition and erosion which have affected this granite rock body (during the past 2.1 Ga) since that time.

While it is acknowledged that fracture propagation is an ongoing process, the propagation curve derived here has an exponential form, suggesting that the Room 209 fracture has been non-propagating or "dormant" for the last several hundred million years. This "dormancy" appears likely to continue so long as this area is remote from deformation at the continental margin or from continental rifting, the sorts of events necessary to radically change the stresses and topography.

5. Acknowledgements

This work was done as part of the Canadian Nuclear Fuel Waste Management Program, which is jointly funded by AECL and Ontario Hydro under the auspices of the CANDU Owners Group.

6. References

Adams, J. and P. Basham. 1991. The seismicity and seismotectonics of eastern Canada. In Neotectonics of North America, D.B. Slemmons, E.R. Engdahl, M.D. Zoback and D.D. Blackwell (Editors). The Geology of North America, Decade Map Volume 1. Geological Society of America, Boulder, Colorado, 261-276.

Atkinson, G.M. 1992. Seismic hazard in northwestern Ontario. Atomic Energy of Canada Limited Technical Record, TR-M-22.

Atkinson, G.M. and R.K. McGuire. 1993. Probability of damaging earthquakes in northwestern Ontario. Atomic Energy of Canada Limited Technical Record, TR-M-23.

Basham, P.W. and M.G. Cajka. 1985. Contemporary seismicity on Northwestern Ontario. Atomic Energy of Canada Limited Technical Report, TR-299.

Bonilla, M. 1970. Surface faulting and related effects. In Earthquake Engineering. R. Wiegel (Editor). Prentice-Hall Inc., Englewood Cliffs, N.J.

Brown, A., N.M. Soonawala, R.A. Everitt and D.C. Kamineni. 1989. Geology and geophysics of Underground Research Laboratory site, Lac du Bonnet, Batholith, Manitoba. Canadian Journal of Earth Sciences, 26 404-425.

Brown, A., R.A. Everitt, C.D. Martin and C.C. Davison. 1994. Past and future fracturing in AECL research areas in the Superior Province of the Canadian precambrian shield; with emphasis on the Lac du Bonnet batholith. Atomic Energy of Canada Limited Report.

Dowding, C.H. and A. Rozen. 1978. Damage to rock tunnels from earthquake shaking. Journal Geotechnical Engineering Division, ASCE, 104, 229-247.

Everitt, R.A. and A. Brown. 1986. Subsurface geology of the Underground Research Laboratory. An overview of recent developments. In Proceedings of the 20th Information Meeting of the Canadian Nuclear Fuel Waste Management Program. 1985. Winnipeg, Manitoba. Atomic Energy of Canada Limited Technical Record, TR-375, 146-181.

Everitt, R.A.., and A. Brown (in press). Geological mapping of AECL Research's Underground Research Laboratory. A cross section of thrust faults and associated fractures in the roof zone of an Archean batholith. In L.R. Myer (editor) Pre-prints of International Symposium on Fractured and Jointed Rock Masses, a regional conference of the International Society for Rock Mechanics, June 1-6th 1992. Granlibakken, California. Volume 1, 1-11.

Everitt, R.A., A. Brown, C.C. Davison, M. Gascoyne, C.D. Martin. 1990. Regional and local setting of the Underground Research Laboratory. In R.S. Sinha (editor) Proceedings of the International Symposium on Unique Underground Structures. July 1990. Denver, Colorado.

Everitt, R.A., P. Gann, and D.M. Boychuk. 1993. Mine-by experiment: Part 7 - geological setting and general geology. Atomic Energy of Canada Limited Report, RC-1080.1.

Gascoyne, M. and J.J. Cramer. 1987. History of actinide and minor element mobility in an archean granite batholith in Manitoba. Applied Geoscience 2, (1) 37-53.

Lang, P.A., R.A. Everitt, E.T. Kozak, and C.C. Davison. 1988. Underground Research Laboratory Room 209 instrument array: Pre-excavation information for modellers. Atomic Energy of Canada Limited Report, AECL-9566-1.

Martin, C.D., N.A. Chandler and A. Brown. 1994. Evaluating the potential for large-scale fracturing at a disposal vault: an example using the Underground Research Laboratory. Atomic Energy of Canada Limited. Report AECL-11180.

Stone, D. and D.C. Kamineni. 1982. Fractures and fracture infillings of the Eye Dashwa Lakes pluton, Atikokan, Ontario. Canadian Journal of Earth Science 19, 789-803.

Stone, D., D.C. Kamineni, A. Brown, and R.A. Everitt. 1989. A comparison of fracture styles in two granite bodies of the Superior Province. Canadian Journal of Earth Sciences, 26, 387-403.

Syme, E.C., W. Weber and P.G. Lenton. 1993. Manitoba geochronology database. Manitoba Energy and Mines Open File Report #OF93-4.

Wyss, M. 1979. Estimating maximum expectable magnitude of earthquakes from fault dimensions. Geology, 7, 336-340.

Internal reports, available from SDDO, AECL Research, Chalk River Laboratories, Chalk River, Ontario, K0J 1J0.

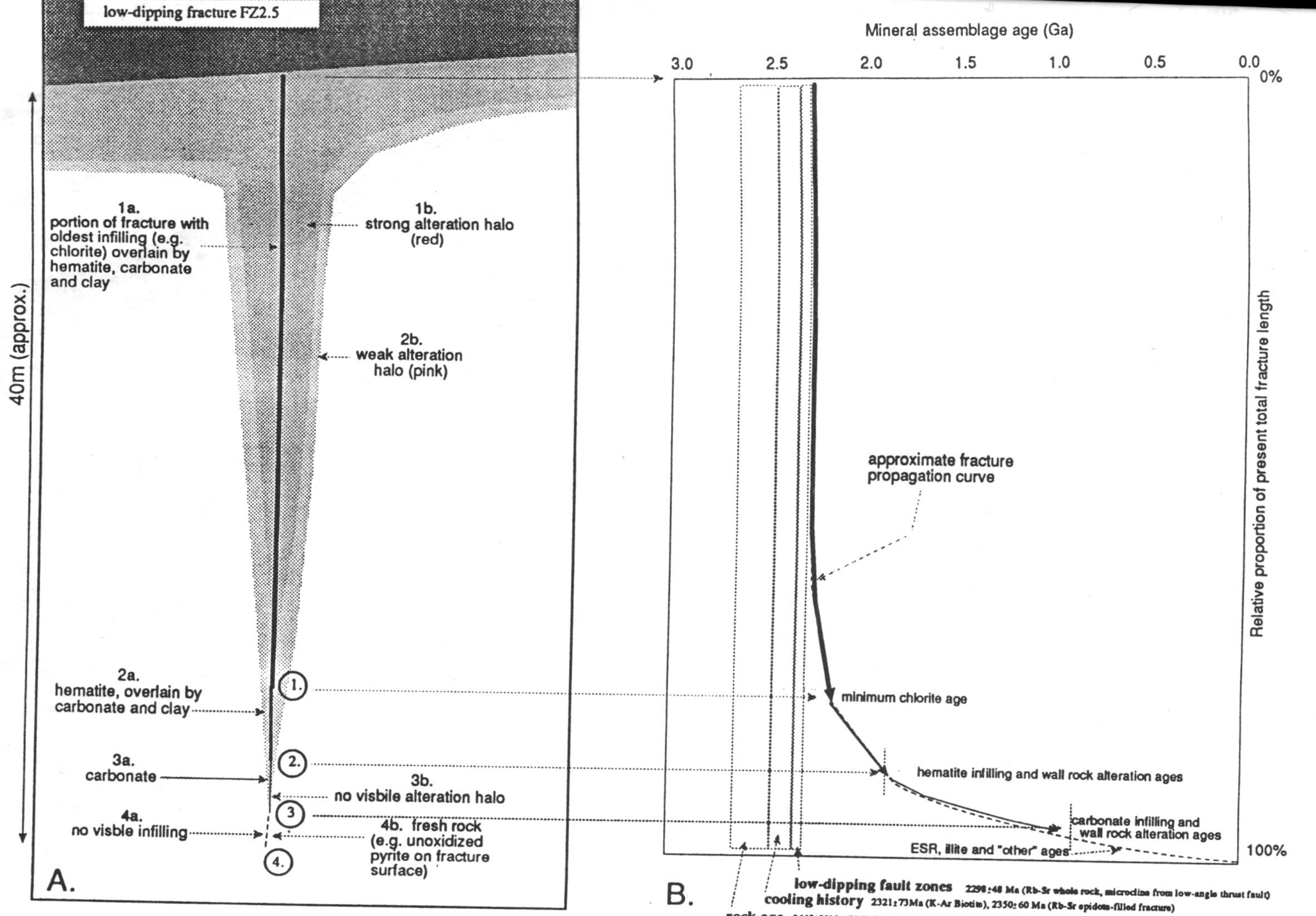

Figure 3: Simplified cross section of the Room 209 Fracture (A) showing the distribution of the infilling and wall rock alteration, and the propagation curve (B) showing the respective ages for the infillings and alterations.

Table 1: Summary of age data:

CRYSTALLIZATION:[1,2,3]		
	2670$^+$/.2470 Ma	
	2665$^+$/.20 Ma	U-Pb zircon[1]
	2603$^+$/. 97 Ma	Rb-Sr whole rock 1
	2571$^+$/. 33 Ma	RB-Sr whole rock 1
DYKE INTRUSION AND EARLY FRACTURING:[1]		
early granitic segregation	2475? Ma est.	
granodiorite dykes	2450? Ma est.	
pegmatite and aplite dykes	2400? Ma est.	
DEUTERIC ALTERATION, RECRYSTALLIZATION AND FRACTURING:[1] **2470-2100 Ma**		
	2475 Ma	Rb-SrK-feldspar[1]
	2365$^+$/.02 Ma	Ar-Ar biotite, cooling at blocking temperature of 300°C
	2321$^+$/.71 Ma	K-Ar biotite[1]
epidotized fracture	2350+60	Rb-Sr[1]
low angle thrust fault	2298+48	microcline-whole rock[1,3]
HYDROTHERMAL (METEORIC) ALTERATION:2100-470 Ma. Formation of illite, and some hematite and carbonate		
illite in fracture zone 2	832$^+$/.	(Ar-ar)biotite[1,3] illite age formed by breakdown of chlorite to illite and hematite
illite in fracture zone 3	722$^+$/.3	(Ar-Ar)biotite[1,3](age probably corresponds with 832 Ma or 510 Ma of fractures in FZ2
illite in fracture zone 2	510$^+$/.	(Ar-Ar)biotite[1,3](overpreinting of illite possibly associated with downwarping
		and initiation of Ordovician carbonate, sedimentation, suggested by post-clay carbonate
infillings of fractures.		
LOW-TEMPERATURE HYDROTHERMAL: 470 Ma - present:[1,2,3]		

[1] Brown et al 1989
[2] Gascoyne and Cramer 1987
[3] Brown et al 1994

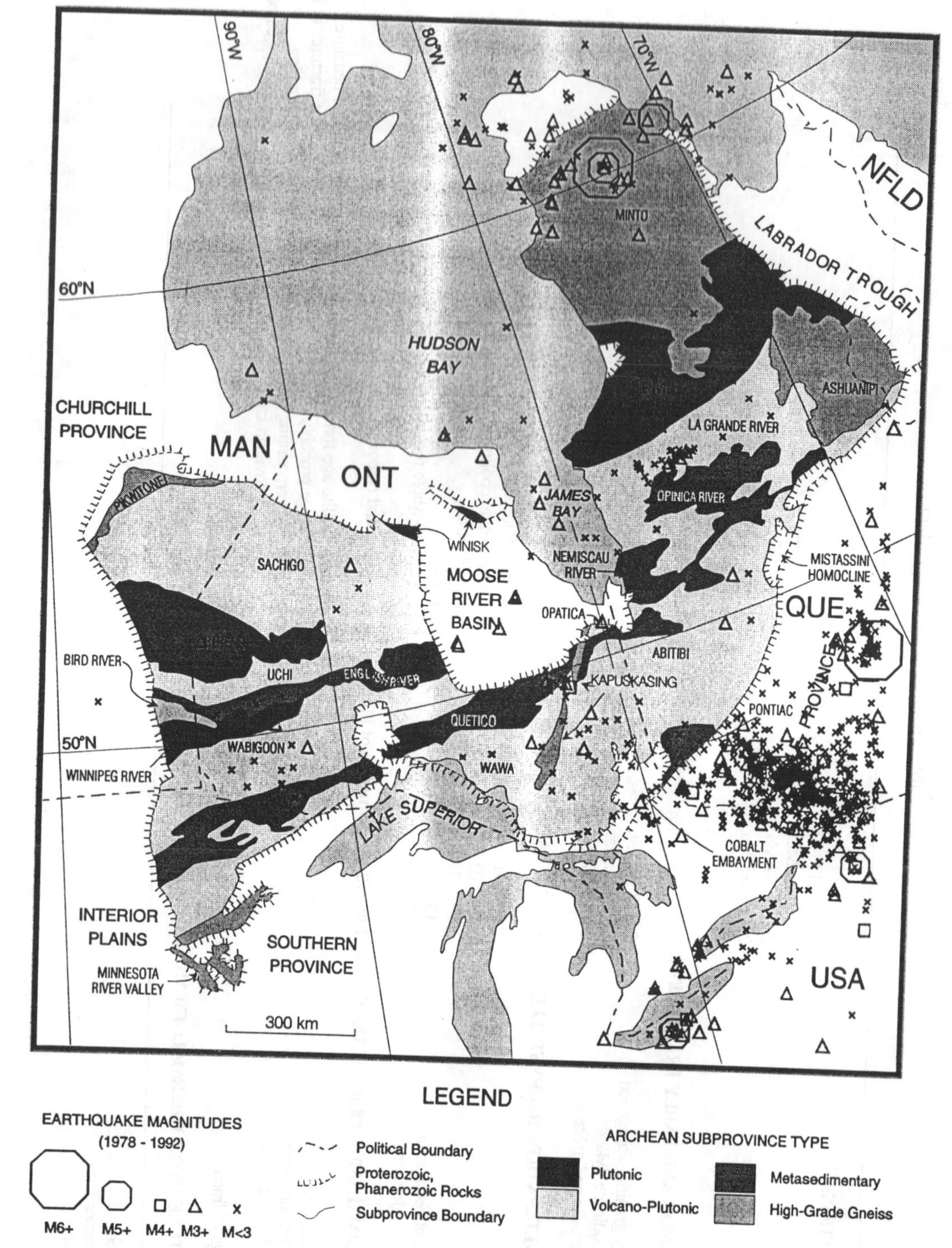

Figure 1: Seismicity on the Archean Canadian Shield

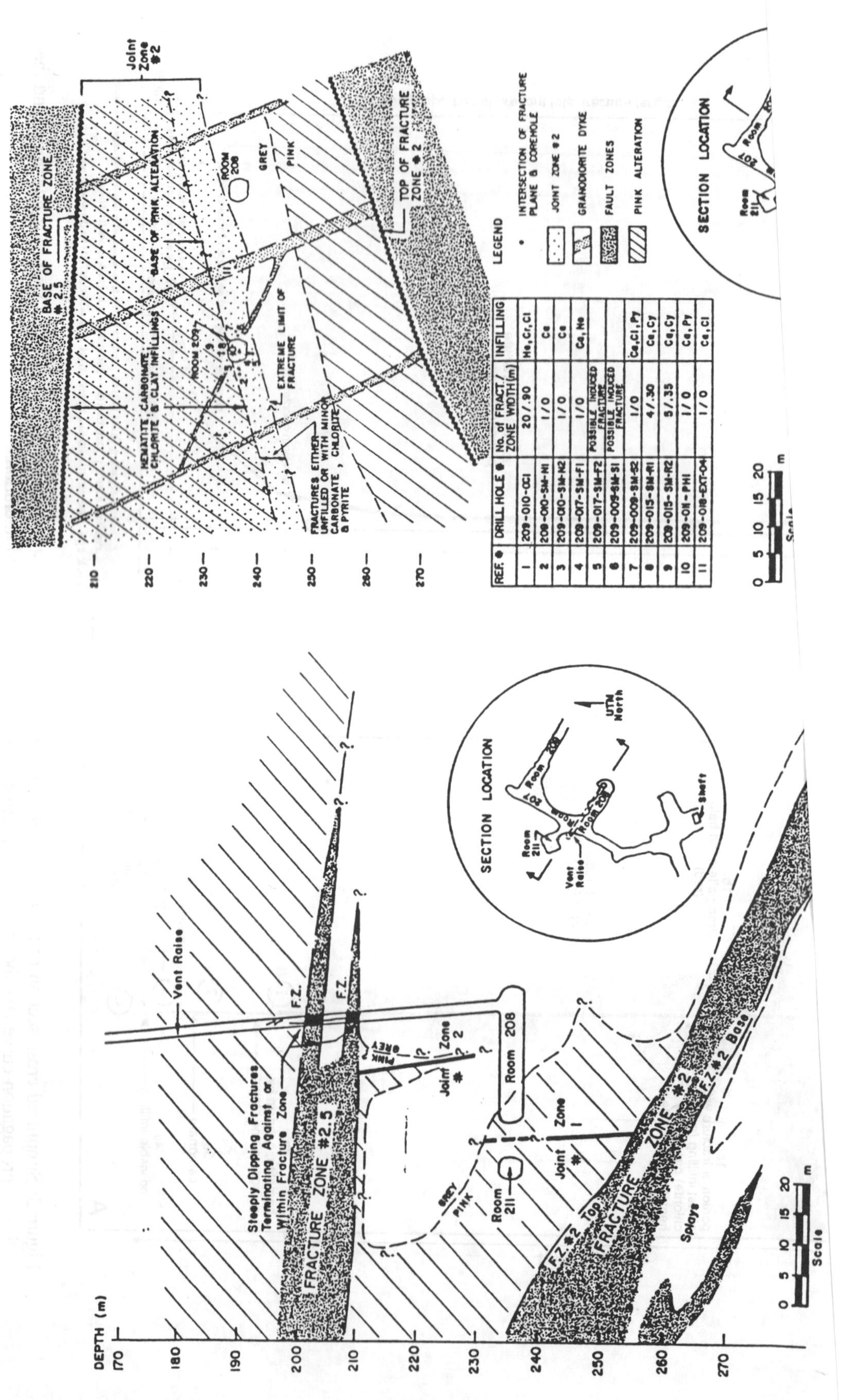

REF. #	DRILL HOLE #	No. of FRACT./ ZONE WIDTH (m)	INFILLING
1	209-010-OC1	20/.90	He,Cr,Cl
2	209-010-SM-N1	1/0	Ca
3	209-010-SM-N2	1/0	Ca
4	209-017-SM-F1	1/0	Ca, He
5	209-017-SM-F2	POSSIBLE INDUCED FRACTURE	
6	209-009-SM-S1	POSSIBLE INDUCED FRACTURE	
7	209-009-SM-S2	1/0	Ca,Cl,Py
8	209-015-SM-R1	4/.30	Ca,Cy
9	209-015-SM-R2	5/.35	Ca,Cy
10	209-011-PH1	1/0	Ca,Py
11	209-018-EXT-04	1/0	Ca,Cl

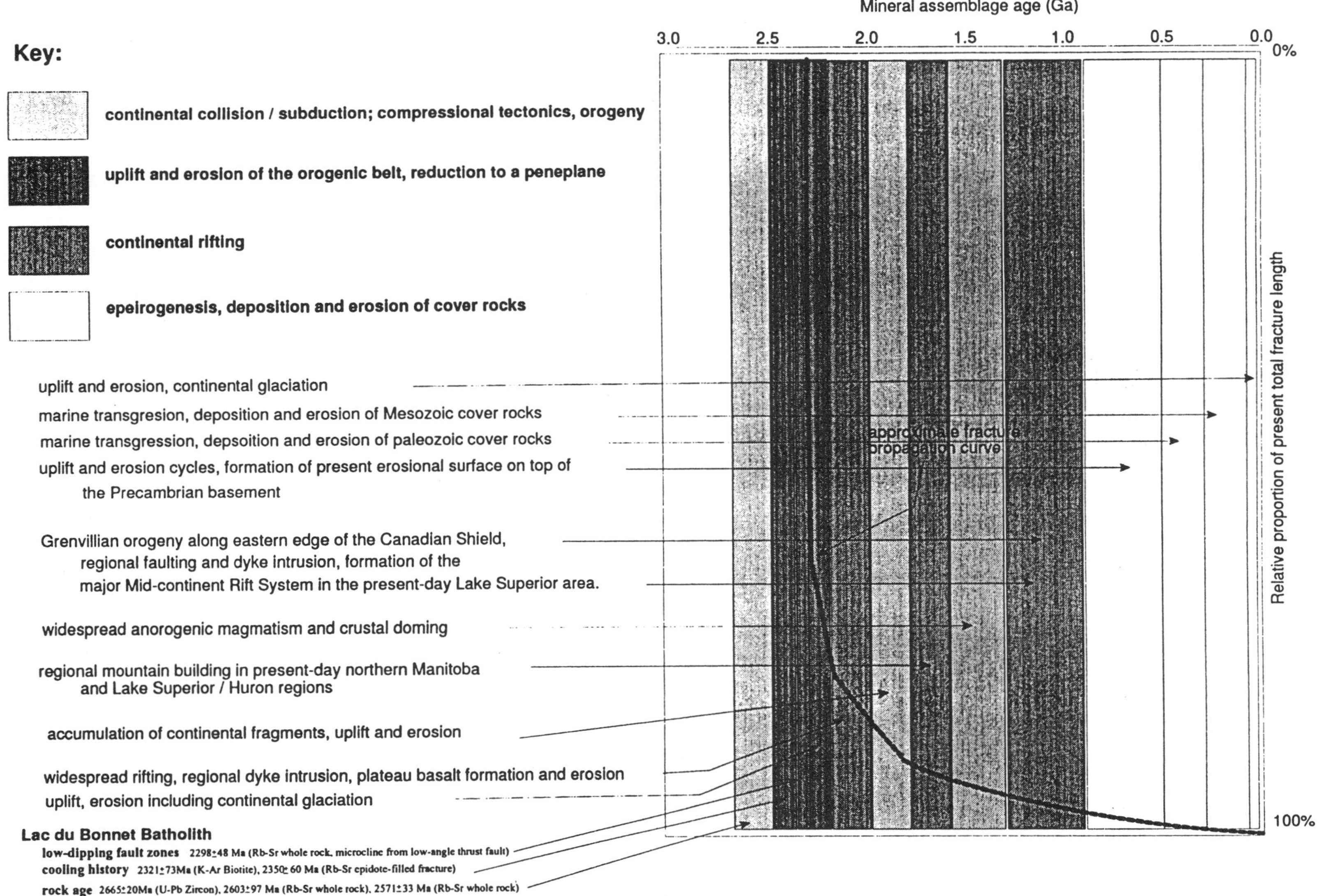

Figure 4: The fracture propagation curve from Figure 3, superimposed on a generalized summary of the regional tectonic events for this area. Four general types of events are distinguished by the grey shading, while specific tectonic events are also listed.

The Role of Long-Term Geologic Changes in the Regulation of the Canadian Nuclear Fuel Waste Management Program

P. Flavelle
Atomic Energy Control Board
Ottawa, Canada

Abstract

It is recognized that the geosphere is a dynamic system over the long time frames of nuclear fuel waste disposal. This paper describes how consideration of a dynamic geosphere has impacted upon the evolving regulatory environment in Canada, and how the approach taken to comply with the regulatory requirements can affect the evaluation of long-term geologic changes. AECB staff opinion is that if the maximum possible effect of geologic changes can be demonstrated to have negligible impact on the safety of a nuclear fuel waste repository, then further consideration of a dynamic geosphere is unnecessary for the current review of the Canadian Nuclear Fuel Waste Management Program.

Résumé

Il est bien connu que la géosphère, dans le contexte du stockage permanent des déchets de combustible nucléaire, est un système dynamique à long terme. Cette communication décrit comment le caractère dynamique de la géosphère a influencé les exigences réglementaires au Canada, et vice-versa, comment ces dernières vont influencer la façon d'évaluer les changements géologiques à long terme de la géosphère. Dans l'opinion du personnel de la CCEA, si on peut démontrer que l'effet le plus grand possible des changements géologique ne produit que peu de risque sur la sécurité d'une installation de stockage permanent, alors aucune considération additionnelle de la géosphère dynamique n'est nécessaire pour l'étape présente d'examen du Programme canadien de gestion des déchets de combustible nucléaire.

Background: The regulatory environment and long-term geologic changes

The Canadian Nuclear Fuel Waste Management Program (CNFWMP) was initiated by government decree in 1978 [1]. The program is the responsibility of Atomic Energy of Canada Limited (AECL) and Canada's largest nuclear power generator, Ontario Hydro. The process for evaluating the program was announced in 1981 [2]. The Atomic Energy Control Board (AECB), the Canadian nuclear regulatory authority, would provide the regulatory guidance for AECL to produce a Concept Assessment Document to "assess whether permanent disposal in a deep underground repository in intrusive igneous rock is a safe, secure and desirable method of disposing of nuclear fuel waste".

The guidelines and regulatory requirements for evaluating the nuclear fuel waste disposal concept are contained in Regulatory Policy Statement R-71 [3]. Although the time scale for evaluating the concept is not specified in R-71, it is alluded to in such phrases as "the system will perform over the long term as required".

In 1981 and in 1985 AECL issued the first and second Interim Concept Assessment Documents [4, 5]. In their review of the first Interim Concept Assessment Document, AECB staff noted that "time evolution of the system and disruptive events" were not included in the assessment of long-term performance. The review identified "the variation of the system with time" as "a modelling aspect of crucial importance in examining the long-term performance of a deep geological disposal facility for nuclear fuel waste."

In the AECB review of the second Interim Concept Assessment Document it was noted that long-term transient conditions from vault construction, operation and closure were not evaluated explicitly. Nor were long-term processes that could affect buffer/backfill evolution included. Evaluation of these would have required knowledge or assumption of the "normal" evolution of a site due to geologic processes and events.

The time scale for the assessment of the long-term post-closure period is addressed in R-104 [6]. Ten thousand years is identified as the limit beyond which it is felt that uncertainties in model scenarios and parameters outweigh any possible accuracy in the predictions, and reporting significant figures becomes meaningless. Nonetheless, "reasoned arguments" must be presented that NFW disposal will not result in "acute radiological risks" to individuals at any time, even if numerical accuracy cannot be achieved in the quantitative predictions of performance.

The "reasoned arguments" can take many forms. They can be based on semi-quantitative or qualitative evaluations of theoretical analyses and/or analogue studies [7]. In doing so, the bounds of facility performance should be evaluated for all credible conditions to demonstrate that no unacceptable risks occur. The credible conditions should not be limited to only a suite of steady-state conditions. Transient conditions, caused by geologic processes, discrete geologic events and any cumulative effects of the processes and events, should also be assessed for unacceptable impacts on repository safety. For example, the rate of advance or retreat of a glacier may have more impact than the static load of the glacier.

Long-term geologic changes and repository safety

Changes with time in the geosphere are driven by disequilibrium in the system due to changes in the external conditions to which the geosphere is subjected. Geologic processes produce a continuous evolution toward equilibrium conditions. Geologic events are discrete occurrences which may cause disequilibrium on a local scale, but on a larger scale also tend toward equilibrium conditions. For this discussion, the distinction between geologic processes and events is irrelevant because the use of both terms is meant to be inclusive. The complete spectrum of processes and events should be considered, including cumulative effects that they may produce. For example, alteration of fracture infilling minerals as the system evolves toward chemical equilibrium may have a cumulative effect on fracture strength, such that a low-magnitude seismic event could have more impact on the fracture aperture in the future than it would today.

It is useful to distinguish between the parameters that geologic processes\events influence directly and those parameters important for safety of the disposal system. The two sets may have parameters in common, and the safety-related parameters may also be subject to secondary or tertiary effects from the primary changes in the geologic parameters, as illustrated in Figure 1. The changes resulting from geologic processes or events (or their cumulative effects) will not be significant if the parameter that is changing initially does not affect repository safety (either directly or by secondary or tertiary changes in safety-related parameters). A performance assessment of a repository should identify the important safety-related parameters, which may include distance to a hydraulically conductive zone, spatial frequency and hydraulic aperture of fractures connecting repository rooms to a conductive zone, normal stress across open fractures and hydraulic gradient.

An example of these possible changes for a seismic event is illustrated in Figure 2. In this example, the seismic event causes a primary change in the in situ stress, which results in a secondary change in the normal stress across a fracture. If this secondary change is not large enough to lead to an irreversible change in the fracture aperture (a tertiary change), then there would be no subsequent effect on the safety of the repository as measured by the maximum radiological impact of the contaminants discharged to the biosphere.

As stated in R-104, the AECB requires that the maximum possible effect of long-term changes in the geosphere on the safety of NFW disposal be evaluated, by examination of the magnitude of the changes and consideration of the effects of the rates of geologic changes on repository safety. It will be necessary to characterize both the changes (for magnitude and rate) and the response of the repository-geosphere system to those changes. Climate variation, glaciation, tectonism and seismicity should all be included in the phenomena considered. The changes caused by these phenomena can be applied to the dynamic response of the repository-geosphere system, to estimate their impact on the groundwater flow system and chemistry in the normal evolution of the geosphere at the site.

In addition, the repository itself will affect the normal evolution of the site. The perturbations caused by repository construction and operation will tend to increase the disequilibrium which is the driving force for geologic changes. The dynamic response of the system to different perturbations will need to be assessed to determine the magnitude of the deviation from normal evolution and the length of time it will take the system to recover.

The system's dynamic response to the perturbations caused by some of the geologic processes and events may be derived as part of the site characterization, but its response to other perturbations will likely have to be inferred from paleohydrogeologic and paleohydrogeochemical studies, as well as from natural analogues and any other examinations of historic long-term geologic change. For example, the response of the flow system to changes in stress might be measured directly or inferred from direct measurements of physical properties. However, the response of the flow system's contaminant sorption capacity to the introduction of foreign water may have to be inferred from paleohydrogeochemical studies of the evolution of fracture infilling minerals. The response to stress changes would be of use in evaluating the effects of seismicity, repository excavation and glacial loading, while the response to foreign water is needed to predict the impact of introducing surface water and buffer\backfill porewater into the geosphere at depth.

Satisfying the regulatory requirements regarding long-term geologic changes

There are two approaches to evaluating the maximum possible effect of long-term changes in the geosphere on the safety of NFW disposal . One way, illustrated in Figure 3, is to predict the maximum possible changes and rates of changes in geosphere parameters and predict all impacts (primary, secondary or tertiary) on the parameters that affect repository safety. This is similar to an event tree analysis.

In the example presented in Figure 2, the maximum change of in situ stress due to the distribution of magnitudes and distances to epicentre of seismic events would be predicted (primary change). The resulting maximum change in the normal stress across the fractures in the geosphere at the site would be evaluated (maximum secondary changes), as would the associated tertiary effects on fracture aperture, groundwater velocity and subsequent contaminant discharge rates from the geosphere. The overall impact of the maximum possible seismic event on repository safety can then be calculated. In performing these evaluations, it will be important to incorporate cumulative effects when estimating the maximum effects. For example, the response of the fracture aperture to changes in normal stress may evolve with time due to other geologic processes (such as mineral alteration affecting fracture strength).

The alternate approach, illustrated in Figure 4, would be to use the regulatory criteria to estimate the minimum changes in safety-related parameters that would result in unacceptable dose (i.e. system "failure"). The changes in geosphere parameters that would be needed to cause such "failure" would be derived from the changes in safety-related parameters. The phenomena or combinations of phenomena that would cause the geologic changes would be identified and evaluated. This is similar to a fault tree analysis, where the results are the set of initiating conditions or events which could lead to a "failure". As before, consideration of the cumulative effects of processes and events should be included.

For example, if a 'w'% increase in the groundwater velocity would result in an unacceptable dose, it might arise from any number of combinations of geologic processes or events. These might include reactivation of closed fractures by seismic events of magnitude 'x' or greater, a 'y'% increase in hydraulic gradient due to changes in surface water drainage patterns caused by climatic changes, or a 'z'% increase in hydraulic conductivity due to dissolution if fracture infilling minerals by foreign groundwater that becomes part of a thermal convection cell. The magnitudes and rates of the phenomena that would be needed to cause any of these conditions to occur could then be compared to the maximum possible magnitudes and rates of those phenomena.

If there are no unacceptable impacts from the maximum possible long-term geologic changes predicted by the first approach, or if the possible range of long-term geologic changes does not include the minimum needed for an unacceptable impact predicted in the second approach, then a prima facie argument for long-term safety could be made and further consideration of long-term geologic change would be unnecessary. If a prima facie argument for long-term safety cannot be made, then more detailed consideration of the phenomena and their impacts would be necessary. This may involve the inclusion of probability of occurrence of the phenomena, or supporting reasoned arguments such as natural analogues. In either approach to evaluating the effects of long-term geologic changes, it will be necessary to characterize the geologic changes, define the system's response to perturbations, identify the parameters that control repository performance and apply the geologic changes to the system's response to predict how the parameters that control repository performance, and hence the safety of the repository, will be affected.

Summary

The regulatory position adopted by the AECB is that reasoned arguments should be used to demonstrate the safety of a nuclear fuel waste disposal facility beyond the period for quantitative performance assessment of ten thousand years. Such reasoned arguments should consider long-term geologic changes in the bounding evaluation of the system's response to perturbations.

It is the opinion of AECB staff that it should first be shown that long-term geologic changes could have an unacceptable impact on repository safety before more detailed evaluations of the phenomena causing the changes is justified. If bounding calculations show that the maximum possible impact on repository safety is negligible, then in our opinion there is no need for further consideration of long-term geologic changes.

References

1 Energy, Mines and Resources Canada, News Release, "Canada/Ontario Radioactive Waste Management Program", Ottawa (1978)

2 Energy, Mines and Resources Canada, News Release, "Canada-Ontario Joint Statement on the Nuclear Fuel Waste Management Program", Ottawa (1981)

3 Atomic Energy Control Board, "Deep Geological Disposal of Nuclear Fuel Waste: Background Information and Regulatory Requirements Regarding the Concept Assessment Phase", Regulatory Document R-71, Regulatory Policy Statement, Ottawa (1985)

4 Lyon, R.B. et al, "Environmental and Safety Assessment Studies for Nuclear Fuel Waste Management", Atomic Energy of Canada Limited Technical Record TR-127 (volumes 1-3), Pinawa Manitoba (1981)*

5 Wuschke, D.M et al, "Second Interim Assessment of the Canadian Concept for Nuclear Fuel Waste Disposal", Atomic Energy of Canada Limited Report AECL-8373 (volumes 1-4), Pinawa Manitoba (1985)

6 Atomic Energy Control Board, "Regulatory Objectives, Requirements and Guidelines for the Disposal of Radioactive Wastes - Long-Term Aspects", Regulatory Document R-104, Regulatory Policy Statement, Ottawa (1987)

7 Flavelle, Peter A., "Regulatory Perspectives of Concept Assessment", in DOE/AECL '87: Proceedings of the Conference on Geostatistical, Sensitivity, and Uncertainty Methods for Groundwater Flow and Radionuclide Transport Modeling, San Francisco 1987, Bruce E. Buxton (editor), Batelle Press, Columbus, Ohio (1989); also Atomic Energy Control Board Information Document Info-0256, Ottawa (1987).

* Unrestricted, unpublished report available from SDDO, Atomic Energy of Canada Research Company, Chalk River, Ontario, K0J 1J0

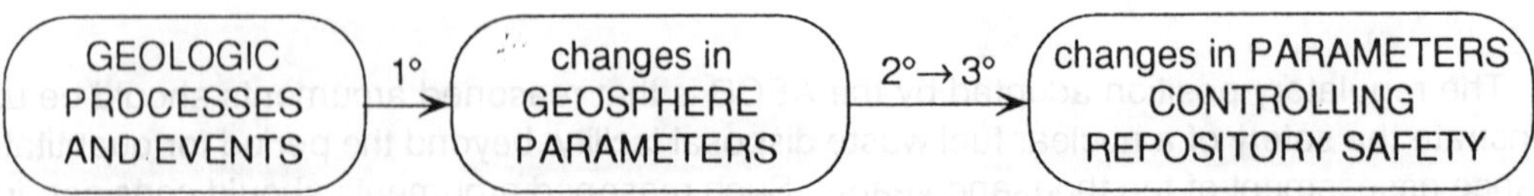

Figure 1. **The distinction between parameters that initially undergo long-term geologic changes due to geologic processes and events, and those that undergo secondary and tertiary changes due to their relationships with the parameters that change initially.**

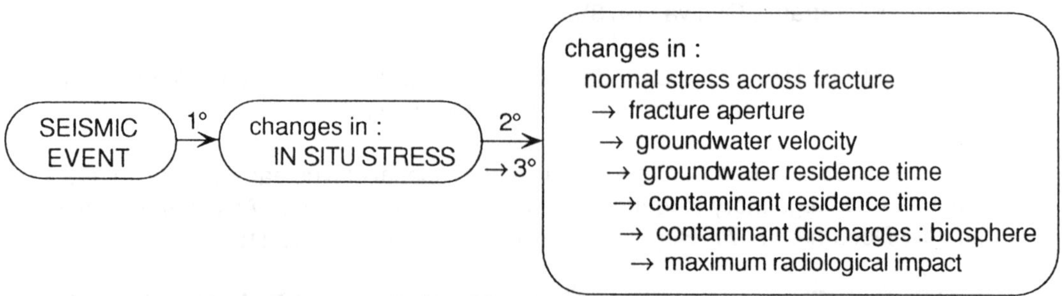

Figure 2. **One example of the primary, secondary and tertiary effects of a seismic event. Changes in the groundwater flow system due to tertiary effects may degrade the long- term performance of the repository.**

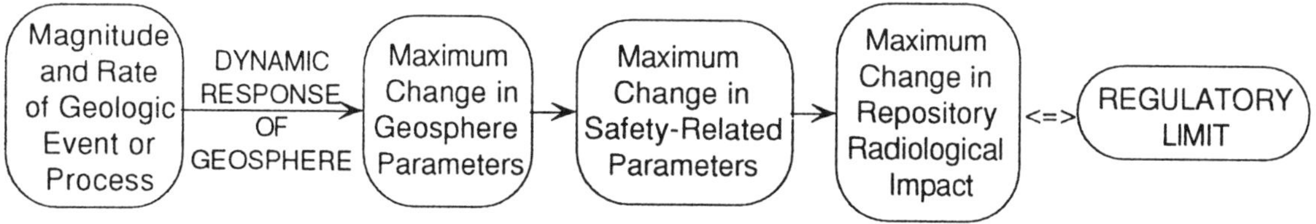

Figure 3. **An approach to evaluating the impact of long-term geologic changes**

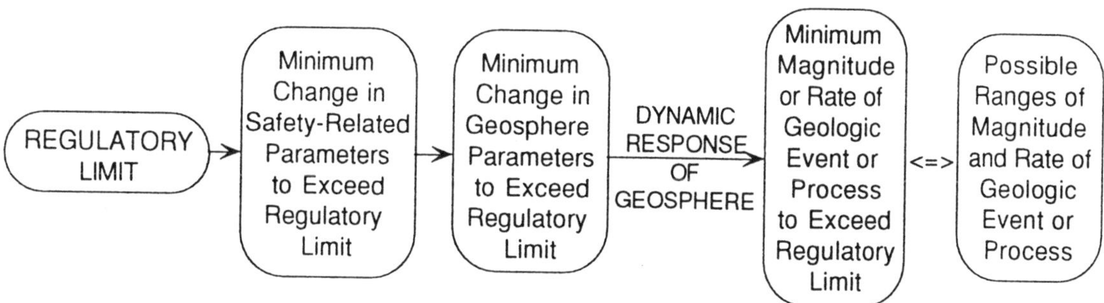

Figure 4. **An alternate approach to evaluating the impact of long-term geologic changes**

SEDE Workshop on

Characterisation of Long-Term Geological Changes Being Considered for Disposal Sites

19th-21st September 1994
Paris

LIST OF PARTICIPANTS (*)

BELGIUM

Isabelle WEMAERE
Waste & Disposal Unit
SCK/CEN
Boeretang 200
B-2400 Mol
Tel: +32 (14) 33 37 70
Fax: +32 (14) 32 12 79

CANADA

Clifford C. DAVISON
Applied Geoscience Branch
AECL Research
Whiteshell Laboratories
Pinawa, Manitoba, R0E 1L0
Tel: +1 (204) 753 2311 x 2299
Fax: +1 (204) 753 2455

Peter A. FLAVELLE
Wastes and Impacts Division
Atomic Energy Control Board
P.O. Box 1046, Station "B"
280 Slater Street
Ottawa, Ontario, K1P 5S9
Tel: +1 (613) 995 3816
Fax: +1 (613) 995 5086

FINLAND

Timo ÄIKÄS
Nuclear Waste Management
Teollisuuden Voima Oy
Annankatu 42 C
SF-00100 Helsinki
Tel: +358 (0) 6180 3730
Fax: +358 (0) 6180 2570

(*) Participant's particulars at the time of the Workshop

Paavo VUORELA | Tel: +358 (0) 46931
Geological Survey of Finland | Fax: +358 (0) 462 205
Betonimiehenkuja 4
SF-02150 Espoo

FRANCE

Jean-François ARANYOSSY | Tel: +(1) 41 17 82 82
ANDRA | Fax: +(1) 41 17 84 08
BP 38
92266 Fontenay-aux-Roses Cedex

Antoine AUGUSTIN | Tel: +(1) 41 17 84 13
ANDRA - Géoprospective | Fax: +(1) 41 17 84 08
BP 38
92266 Fontenay-aux-Roses Cedex

Jean-Yves BOISSON | Tel: +(1) 46 54 80 73
Commissariat à l'Energie Atomique (CEA) | Fax: +(1) 47 35 14 23
Institut de Protection et de Sûreté Nucléaire (IPSN)
BP 6
92265 Fontenay-aux-Roses Cedex

GERMANY

H.-J. HERBERT | Tel: +49 (531) 8012 250
GSF-Forschungszentrum für Umwelt und Gesundheit GmbH | Fax: +49 (531) 8012 200
Institut für Tieflagerung
Postfach 2163
D-38011 Braunschweig

Manfred WALLNER | Tel: +49 (511) 643 2422
Bundesanstalt für Geowissenschaften und Rohstoffe (BGR) | Fax: +49 (511) 643 2304
Stilleweg 2
D-30655 Hannover

JAPAN

Yasuhisa YUSA
General Manager
Geological Environment
Research Section
Tono Geoscience Center
Power Reactor and Nuclear Fuel
Development Corporation (PNC)
959-31 Jorinji, Izumi-cho, Toki, Gifu

Tel: +81 (572) 53 0211
Fax: +81 (572) 55 0180

SPAIN

Carlos del OLMO
ENRESA
Calle Emilio Vargas, 7
E-28043 Madrid

Tel: +34 (1) 519 5206
Fax: +34 (1) 519 5268

SWEDEN

Lars O. ERICSSON
Manager of Geoscience
Swedish Nuclear Fuel and Waste
Management Co. (SKB)
Box 5864
S-102 40 Stockholm

Tel: +46 (8) 665 2800
Fax: +46 (8) 661 5719

Fritz KAUTSKY
Swedish Nuclear Power Inspectorate
(SKI)
Box 27106
S-102 52 Stockholm

Tel: +46 (8) 665 4400
Fax: +46 (8) 661 9086

SWITZERLAND

Peter BITTERLI
Nuclear Safety Inspectorate
HSK
CH-5232 Villigen-HSK

Tel: +41 (56) 99 39 92
Fax: +41 (56) 99 39 07

Marc THURY
NAGRA
Hardstrasse 73
CH-5430 Wettingen

Tel: +41 (56) 37 11 11
Fax: +41 (56) 37 12 07

UNITED KINGDOM

Allan ASHWORTH
HMIP (DOE)
2, Marsham Street
London SW1P 3EB

Tel: +44 (71) 276 4752
Fax: +44 (71) 276 6544

Robert CHAPLOW
UK Nirex Ltd
Curie Avenue
Harwell, Didcot
Oxfordshire OX11 0RH

Tel: +44 (235) 825 600
Fax: +44 (235) 825 624

Alan J. HOOPER
UK Nirex Ltd
Curie Avenue
Harwell, Didcot
Oxfordshire OX11 0RH

Tel: +44 (235) 825 500 x 401
Fax: +44 (235) 820 560

Tim McEWEN
Intera Information Technologies Ltd
Environmental Division
Geosciences Group
47 Burton Street, Melton Mowbray
Leicestershire LE13 1AF

Tel: +44 (664) 411445
Fax: +44 (664) 411402

UNITED STATES OF AMERICA

Robert A. BILLS
USDOE/WIPP
P.O. Box 3090
Carlsbad, NM 88220

Tel: +1 (505) 234 7481
Fax : +1 (505) 234 7430

Dwight T. HOXIE
U.S. Geological Survey
Suite 860, 101 Convention Center Drive
Las Vegas, NV 89109

Tel: +1 (702) 794 7286
Fax: +1 (702) 794 7090

Allen R. LAPPIN
Sandia National Laboratories
P.O. Box 5800, MS 1345
Organization 6307
Albuquerque, NM 87185-1345

Tel: +1 (505) 848 0782
Fax: +1 (505) 848 0789

Linda L. LEHMAN
L. Lehman & Associates, Inc.
1103 W. Burnsville Parkway
Suite 209
Burnsville, MN 55337
Tel: +1 (612) 894 0357
Fax: +1 (612) 894 5028

Robert A. LEVICH
Yucca Mountain Site Characterization Office
U.S. Department of Energy
Post Office Box 98608
Las Vegas, NV 89193
Tel: +1 (702) 794 7946
Fax: +1 (702) 794 7325

J. Timothy SULLIVAN
Yucca Mountain Site Characterization Office
U.S. Department of Energy
Post Office Box 98608
Las Vegas, NV 89193
Tel: +1 (702) 794 7915
Fax : +1 (702) 794 7907

Roger WILMOT
Galson Sciences Ltd
5 Grosvenor House
Melton Road
Oakham, Rutland LE15 6AX
United Kingdom
Tel: +44 (1572) 770649
Fax: +44 (1572) 770650
E-mail: rdw@galson.demon.co.uk

INTERNATIONAL ATOMIC ENERGY AGENCY

Jacques TAMBORINI
Division of Nuclear Fuel Cycle and
Waste Management
International Atomic Energy Agency
P.O. Box 100
A-1400 Vienna
Austria
Tel: +43 (1) 2360 2673
Fax: +43 (1) 234 564

NEA SECRETARIAT

Claudio PESCATORE
Radiation Protection and
Waste Management Division
OECD Nuclear Energy Agency
Le Seine Saint Germain
12, boulevard des Iles
92130 Issy-les-Moulineaux
France
Tel: +(1) 45 24 10 48
Fax: +(1) 45 24 11 10
E-mail: pescatore@nea.fr

MAIN SALES OUTLETS OF OECD PUBLICATIONS
PRINCIPAUX POINTS DE VENTE DES PUBLICATIONS DE L'OCDE

ARGENTINA – ARGENTINE
Carlos Hirsch S.R.L.
Galería Güemes, Florida 165, 4° Piso
1333 Buenos Aires Tel. (1) 331.1787 y 331.2391
Telefax: (1) 331.1787

AUSTRALIA – AUSTRALIE
D.A. Information Services
648 Whitehorse Road, P.O.B 163
Mitcham, Victoria 3132 Tel. (03) 9210.7777
Telefax: (03) 9210.7788

AUSTRIA – AUTRICHE
Gerold & Co.
Graben 31
Wien I Tel. (0222) 533.50.14
Telefax: (0222) 512.47.31.29

BELGIUM – BELGIQUE
Jean De Lannoy
Avenue du Roi 202 Koningslaan
B-1060 Bruxelles Tel. (02) 538.51.69/538.08.41
Telefax: (02) 538.08.41

CANADA
Renouf Publishing Company Ltd.
1294 Algoma Road
Ottawa, ON K1B 3W8 Tel. (613) 741.4333
Telefax: (613) 741.5439

Stores:
61 Sparks Street
Ottawa, ON K1P 5R1 Tel. (613) 238.8985
12 Adelaide Street West
Toronto, ON M5H 1L6 Tel. (416) 363.3171
Telefax: (416)363.59.63

Les Éditions La Liberté Inc.
3020 Chemin Sainte-Foy
Sainte-Foy, PQ G1X 3V6 Tel. (418) 658.3763
Telefax: (418) 658.3763

Federal Publications Inc.
165 University Avenue, Suite 701
Toronto, ON M5H 3B8 Tel. (416) 860.1611
Telefax: (416) 860.1608

Les Publications Fédérales
1185 Université
Montréal, QC H3B 3A7 Tel. (514) 954.1633
Telefax: (514) 954.1635

CHINA – CHINE
China National Publications Import
Export Corporation (CNPIEC)
16 Gongti E. Road, Chaoyang District
P.O. Box 88 or 50
Beijing 100704 PR Tel. (01) 506.6688
Telefax: (01) 506.3101

CHINESE TAIPEI – TAIPEI CHINOIS
Good Faith Worldwide Int'l. Co. Ltd.
9th Floor, No. 118, Sec. 2
Chung Hsiao E. Road
Taipei Tel. (02) 391.7396/391.7397
Telefax: (02) 394.9176

CZECH REPUBLIC – RÉPUBLIQUE TCHÈQUE
Artia Pegas Press Ltd.
Narodni Trida 25
POB 825
111 21 Praha 1 Tel. (2) 242 246 04
Telefax: (2) 242 278 72

DENMARK – DANEMARK
Munksgaard Book and Subscription Service
35, Nørre Søgade, P.O. Box 2148
DK-1016 København K Tel. (33) 12.85.70
Telefax: (33) 12.93.87

EGYPT – ÉGYPTE
Middle East Observer
41 Sherif Street
Cairo Tel. 392.6919
Telefax: 360-6804

FINLAND – FINLANDE
Akateeminen Kirjakauppa
Keskuskatu 1, P.O. Box 128
00100 Helsinki

Subscription Services/Agence d'abonnements :
P.O. Box 23
00371 Helsinki Tel. (358 0) 121 4416
Telefax: (358 0) 121.4450

FRANCE
OECD/OCDE
Mail Orders/Commandes par correspondance :
2, rue André-Pascal
75775 Paris Cedex 16 Tel. (33-1) 45.24.82.00
Telefax: (33-1) 49.10.42.76
Telex: 640048 OCDE

Internet: Compte.PUBSINQ @ oecd.org

Orders via Minitel, France only/
Commandes par Minitel, France exclusivement :
36 15 OCDE

OECD Bookshop/Librairie de l'OCDE :
33, rue Octave-Feuillet
75016 Paris Tel. (33-1) 45.24.81.81
(33-1) 45.24.81.67

Dawson
B.P. 40
91121 Palaiseau Cedex Tel. 69.10.47.00
Telefax: 64.54.83.26

Documentation Française
29, quai Voltaire
75007 Paris Tel. 40.15.70.00

Economica
49, rue Héricart
75015 Paris Tel. 45.78.12.92
Telefax: 40.58.15.70

Gibert Jeune (Droit-Économie)
6, place Saint-Michel
75006 Paris Tel. 43.25.91.19

Librairie du Commerce International
10, avenue d'Iéna
75016 Paris Tel. 40.73.34.60

Librairie Dunod
Université Paris-Dauphine
Place du Maréchal-de-Lattre-de-Tassigny
75016 Paris Tel. 44.05.40.13

Librairie Lavoisier
11, rue Lavoisier
75008 Paris Tel. 42.65.39.95

Librairie des Sciences Politiques
30, rue Saint-Guillaume
75007 Paris Tel. 45.48.36.02

P.U.F.
49, boulevard Saint-Michel
75005 Paris Tel. 43.25.83.40

Librairie de l'Université
12*a*, rue Nazareth
13100 Aix-en-Provence Tel. (16) 42.26.18.08

Documentation Française
165, rue Garibaldi
69003 Lyon Tel. (16) 78.63.32.23

Librairie Decitre
29, place Bellecour
69002 Lyon Tel. (16) 72.40.54.54

Librairie Sauramps
Le Triangle
34967 Montpellier Cedex 2 Tel. (16) 67.58.85.15
Telefax: (16) 67.58.27.36

A la Sorbonne Actual
23, rue de l'Hôtel-des-Postes
06000 Nice Tel. (16) 93.13.77.75
Telefax: (16) 93.80.75.69

GERMANY – ALLEMAGNE
OECD Publications and Information Centre
August-Bebel-Allee 6
D-53175 Bonn Tel. (0228) 959.120
Telefax: (0228) 959.12.17

GREECE – GRÈCE
Librairie Kauffmann
Mavrokordatou 9
106 78 Athens Tel. (01) 32.55.321
Telefax: (01) 32.30.320

HONG-KONG
Swindon Book Co. Ltd.
Astoria Bldg. 3F
34 Ashley Road, Tsimshatsui
Kowloon, Hong Kong Tel. 2376.2062
Telefax: 2376.0685

HUNGARY – HONGRIE
Euro Info Service
Margitsziget, Európa Ház
1138 Budapest Tel. (1) 111.62.16
Telefax: (1) 111.60.61

ICELAND – ISLANDE
Mál Mog Menning
Laugavegi 18, Pósthólf 392
121 Reykjavik Tel. (1) 552.4240
Telefax: (1) 562.3523

INDIA – INDE
Oxford Book and Stationery Co.
Scindia House
New Delhi 110001 Tel. (11) 331.5896/5308
Telefax: (11) 332.5993

17 Park Street
Calcutta 700016 Tel. 240832

INDONESIA – INDONÉSIE
Pdii-Lipi
P.O. Box 4298
Jakarta 12042 Tel. (21) 573.34.67
Telefax: (21) 573.34.67

IRELAND – IRLANDE
Government Supplies Agency
Publications Section
4/5 Harcourt Road
Dublin 2 Tel. 661.31.11
Telefax: 475.27.60

ISRAEL – ISRAËL
Praedicta
5 Shatner Street
P.O. Box 34030
Jerusalem 91430 Tel. (2) 52.84.90/1/2
Telefax: (2) 52.84.93

R.O.Y. International
P.O. Box 13056
Tel Aviv 61130 Tel. (3) 546 1423
Telefax: (3) 546 1442

Palestinian Authority/Middle East:
INDEX Information Services
P.O.B. 19502
Jerusalem Tel. (2) 27.12.19
Telefax: (2) 27.16.34

ITALY – ITALIE
Libreria Commissionaria Sansoni
Via Duca di Calabria 1/1
50125 Firenze Tel. (055) 64.54.15
Telefax: (055) 64.12.57

Via Bartolini 29
20155 Milano Tel. (02) 36.50.83

Editrice e Libreria Herder
Piazza Montecitorio 120
00186 Roma Tel. 679.46.28
Telefax: 678.47.51

Libreria Hoepli
Via Hoepli 5
20121 Milano Tel. (02) 86.54.46
Telefax: (02) 805.28.86

Libreria Scientifica
Dott. Lucio de Biasio 'Aeiou'
Via Coronelli, 6
20146 Milano Tel. (02) 48.95.45.52
Telefax: (02) 48.95.45.48

JAPAN – JAPON
OECD Publications and Information Centre
Landic Akasaka Building
2-3-4 Akasaka, Minato-ku
Tokyo 107 Tel. (81.3) 3586.2016
Telefax: (81.3) 3584.7929

KOREA – CORÉE
Kyobo Book Centre Co. Ltd.
P.O. Box 1658, Kwang Hwa Moon
Seoul Tel. 730.78.91
Telefax: 735.00.30

MALAYSIA – MALAISIE
University of Malaya Bookshop
University of Malaya
P.O. Box 1127, Jalan Pantai Baru
59700 Kuala Lumpur
Malaysia Tel. 756.5000/756.5425
Telefax: 756.3246

MEXICO – MEXIQUE
OECD Publications and Information Centre
Edificio INFOTEC
Av. San Fernando no. 37
Col. Toriello Guerra
Tlalpan C.P. 14050
Mexico D.F.
Tel. (525) 606 00 11 Extension 100
Fax: (525) 606 13 07

Revistas y Periodicos Internacionales S.A. de C.V.
Florencia 57 - 1004
Mexico, D.F. 06600 Tel. 207.81.00
Telefax: 208.39.79

NETHERLANDS – PAYS-BAS
SDU Uitgeverij Plantijnstraat
Externe Fondsen
Postbus 20014
2500 EA's-Gravenhage Tel. (070) 37.89.880
Voor bestellingen: Telefax: (070) 34.75.778

NEW ZEALAND – NOUVELLE-ZÉLANDE
GPLegislation Services
P.O. Box 12418
Thorndon, Wellington Tel. (04) 496.5655
Telefax: (04) 496.5698

NORWAY – NORVÈGE
NIC INFO A/S
Bertrand Narvesens vei 2
P.O. Box 6512 Etterstad
0606 Oslo 6 Tel. (022) 57.33.00
Telefax: (022) 68.19.01

PAKISTAN
Mirza Book Agency
65 Shahrah Quaid-E-Azam
Lahore 54000 Tel. (42) 353.601
Telefax: (42) 231.730

PHILIPPINE – PHILIPPINES
International Booksource Center Inc.
Rm 179/920 Cityland 10 Condo Tower 2
HV dela Costa Ext cor Valero St.
Makati Metro Manila Tel. (632) 817 9676
Telefax: (632) 817 1741

POLAND – POLOGNE
Ars Polona
00-950 Warszawa
Krakowskie Przedmieácie 7 Tel. (22) 264760
Telefax: (22) 268673

PORTUGAL
Livraria Portugal
Rua do Carmo 70-74
Apart. 2681
1200 Lisboa Tel. (01) 347.49.82/5
Telefax: (01) 347.02.64

SINGAPORE – SINGAPOUR
Gower Asia Pacific Pte Ltd.
Golden Wheel Building
41, Kallang Pudding Road, No. 04-03
Singapore 1334 Tel. 741.5166
Telefax: 742.9356

SPAIN – ESPAGNE
Mundi-Prensa Libros S.A.
Castelló 37, Apartado 1223
Madrid 28001 Tel. (91) 431.33.99
Telefax: (91) 575.39.98

Mundi-Prensa Barcelona
Consell de Cent No. 391
08009 – Barcelona Tel. (93) 488.34.92
Telefax: (93) 487.76.59

Llibreria de la Generalitat
Palau Moja
Rambla dels Estudis, 118
08002 – Barcelona
(Subscripcions) Tel. (93) 318.80.12
(Publicacions) Tel. (93) 302.67.23
Telefax: (93) 412.18.54

SRI LANKA
Centre for Policy Research
c/o Colombo Agencies Ltd.
No. 300-304, Galle Road
Colombo 3 Tel. (1) 574240, 573551-2
Telefax: (1) 575394, 510711

SWEDEN – SUÈDE
CE Fritzes AB
S–106 47 Stockholm Tel. (08) 690.90.90
Telefax: (08) 20.50.21

Subscription Agency/Agence d'abonnements :
Wennergren-Williams Info AB
P.O. Box 1305
171 25 Solna Tel. (08) 705.97.50
Telefax: (08) 27.00.71

SWITZERLAND – SUISSE
Maditec S.A. (Books and Periodicals - Livres et périodiques)
Chemin des Palettes 4
Case postale 266
1020 Renens VD 1 Tel. (021) 635.08.65
Telefax: (021) 635.07.80

Librairie Payot S.A.
4, place Pépinet
CP 3212
1002 Lausanne Tel. (021) 320.25.11
Telefax: (021) 320.25.14

Librairie Unilivres
6, rue de Candolle
1205 Genève Tel. (022) 320.26.23
Telefax: (022) 329.73.18

Subscription Agency/Agence d'abonnements :
Dynapresse Marketing S.A.
38, avenue Vibert
1227 Carouge Tel. (022) 308.07.89
Telefax: (022) 308.07.99

See also – Voir aussi :
OECD Publications and Information Centre
August-Bebel-Allee 6
D-53175 Bonn (Germany) Tel. (0228) 959.120
Telefax: (0228) 959.12.17

THAILAND – THAÏLANDE
Suksit Siam Co. Ltd.
113, 115 Fuang Nakhon Rd.
Opp. Wat Rajbopith
Bangkok 10200 Tel. (662) 225.9531/2
Telefax: (662) 222.5188

TUNISIA – TUNISIE
Grande Librairie Spécialisée
Fendri Ali
Avenue Haffouz Imm El-Intilaka
Bloc B 1 Sfax 3000 Tel. (216-4) 296 855
Telefax: (216-4) 298.270

TURKEY – TURQUIE
Kültür Yayinlari Is-Türk Ltd. Sti.
Atatürk Bulvari No. 191/Kat 13
Kavaklidere/Ankara
Tel. (312) 428.11.40 Ext. 2458
Telefax: (312) 417 24 90
Dolmabahce Cad. No. 29
Besiktas/Istanbul Tel. (212) 260 7188

UNITED KINGDOM – ROYAUME-UNI
HMSO
Gen. enquiries Tel. (171) 873 8242
Postal orders only:
P.O. Box 276, London SW8 5DT
Personal Callers HMSO Bookshop
49 High Holborn, London WC1V 6HB
Telefax: (171) 873 8416
Branches at: Belfast, Birmingham, Bristol, Edinburgh, Manchester

UNITED STATES – ÉTATS-UNIS
OECD Publications and Information Center
2001 L Street N.W., Suite 650
Washington, D.C. 20036-4922 Tel. (202) 785.6323
Telefax: (202) 785.0350

Subscriptions to OECD periodicals may also be placed through main subscription agencies.

Les abonnements aux publications périodiques de l'OCDE peuvent être souscrits auprès des principales agences d'abonnement.

Orders and inquiries from countries where Distributors have not yet been appointed should be sent to: OECD Publications Service, 2, rue André-Pascal, 75775 Paris Cedex 16, France.

Les commandes provenant de pays où l'OCDE n'a pas encore désigné de distributeur peuvent être adressées à : OCDE, Service des Publications, 2, rue André-Pascal, 75775 Paris Cedex 16, France.

1-1996

OECD PUBLICATIONS, 2, rue André-Pascal, 75775 PARIS CEDEX 16
PRINTED IN FRANCE
(66 96 06 1) ISBN 92-64-14829-9 – No. 48676 1996

Table 1: Summary of age data:

CRYSTALLIZATION:[1,2,3]		
	2670$^{+}/_{-}$2470 Ma	
	2665$^{+}/_{-}$20 Ma	U-Pb zircon[1]
	2603$^{+}/_{-}$ 97 Ma	Rb-Sr whole rock 1
	2571$^{+}/_{-}$ 33 Ma	RB-Sr whole rock 1
DYKE INTRUSION AND EARLY FRACTURING:[1]		
early granitic segregation	2475? Ma est.	
granodiorite dykes	2450? Ma est.	
pegmatite and aplite dykes	2400? Ma est.	
DEUTERIC ALTERATION, RECRYSTALLIZATION AND FRACTURING:[1] 2470-2100 Ma		
	2475 Ma	Rb-SrK-feldspar[1]
	2365$^{+}/_{-}$02 Ma	Ar-Ar biotite, cooling at blocking temperature of 300^{0}C
	2321$^{+}/_{-}$71 Ma	K-Ar biotite[1]
epidotized fracture	2350+60	Rb-Sr[1]
low angle thrust fault	2298+48	microcline-whole rock[1,3]
HYDROTHERMAL (METEORIC) ALTERATION:2100-470 Ma. Formation of illite, and some hematite and carbonate		
illite in fracture zone 2	832$^{+}/_{-}$	(Ar-ar)biotite[1,3] illite age formed by breakdown of chlorite to illite and hematite
illite in fracture zone 3	722$^{+}/_{-}$3	(Ar-Ar)biotite[1,3](age probably corresponds with 832 Ma or 510 Ma of fractures in FZ2
illite in fracture zone 2	510$^{+}/_{-}$	(Ar-Ar)biotite[1,3](overpreinting of illite possibly associated with downwarping and initiation of Ordovician carbonate, sedimentation, suggested by post-clay carbonate
infillings of fractures.		
LOW-TEMPERATURE HYDROTHERMAL: 470 Ma - present:[1,2,3]		

1. Brown et al 1989
2. Gascoyne and Cramer 1987
3. Brown et al 1994

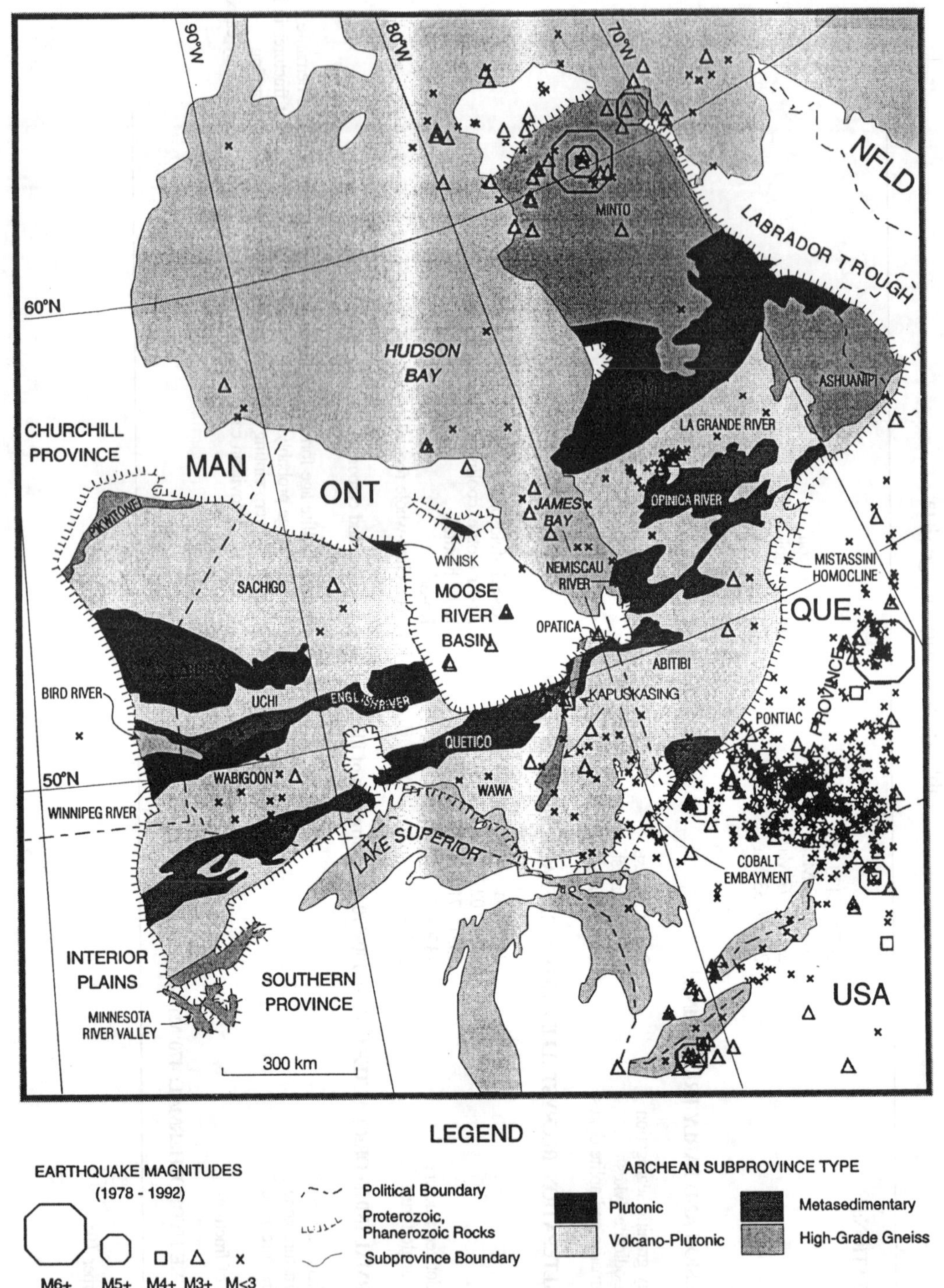

Figure 1: Seismicity on the Archean Canadian Shield

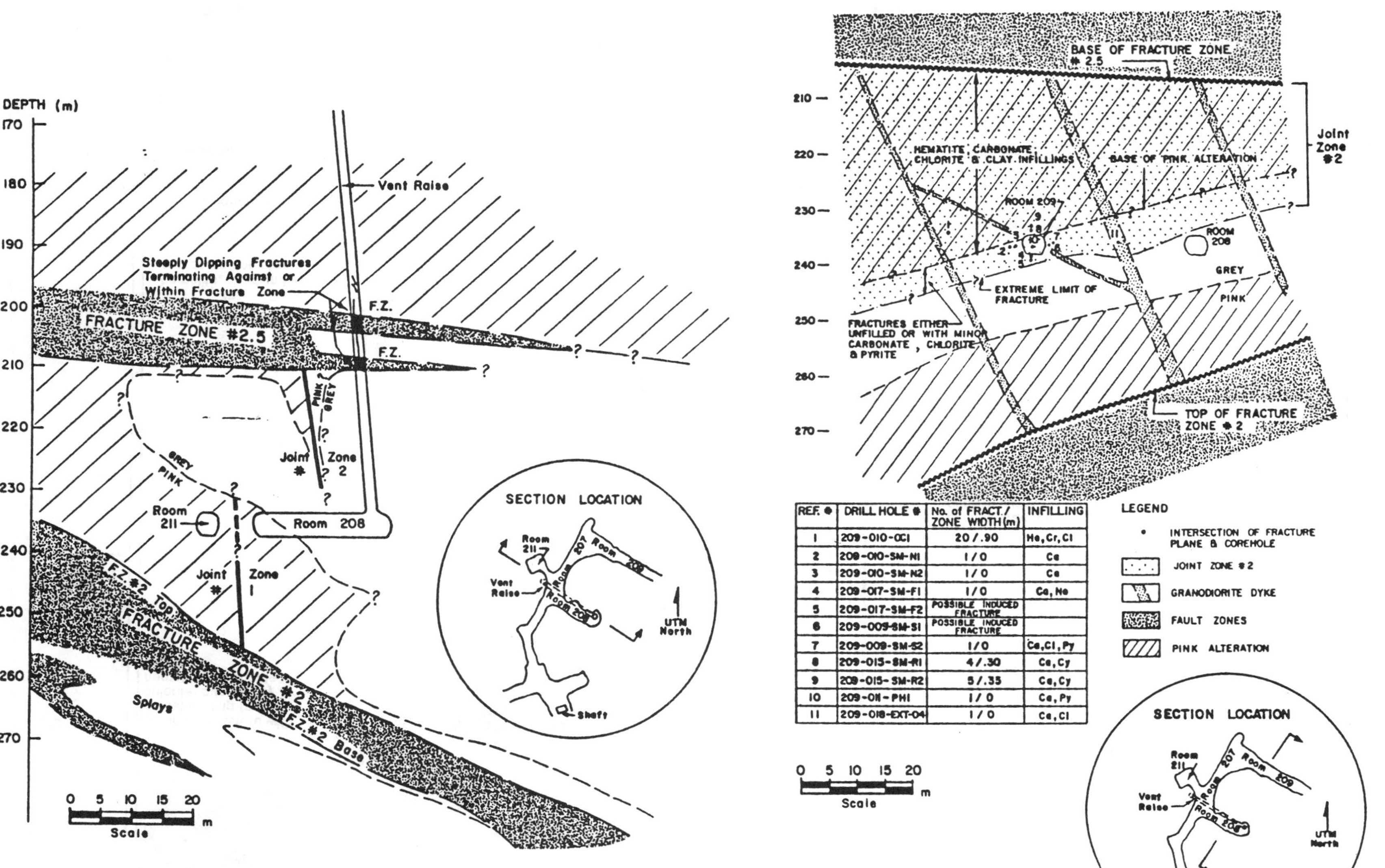

REF. #	DRILL HOLE #	No. of FRACT./ ZONE WIDTH (m)	INFILLING
1	209-010-CC1	20/.90	He, Cr, Cl
2	209-010-SM-N1	1/0	Ca
3	209-010-SM-N2	1/0	Ca
4	209-017-SM-F1	1/0	Ca, He
5	209-017-SM-F2	POSSIBLE INDUCED FRACTURE	
6	209-009-SM-S1	POSSIBLE INDUCED FRACTURE	
7	209-009-SM-S2	1/0	Ca, Cl, Py
8	209-015-SM-R1	4/.30	Ca, Cy
9	209-015-SM-R2	5/.35	Ca, Cy
10	209-011-PH1	1/0	Ca, Py
11	209-018-EXT-04	1/0	Ca, Cl

Figure 2: Sections normal to (left) and in the plane of (right) the Room 209 Fracture.

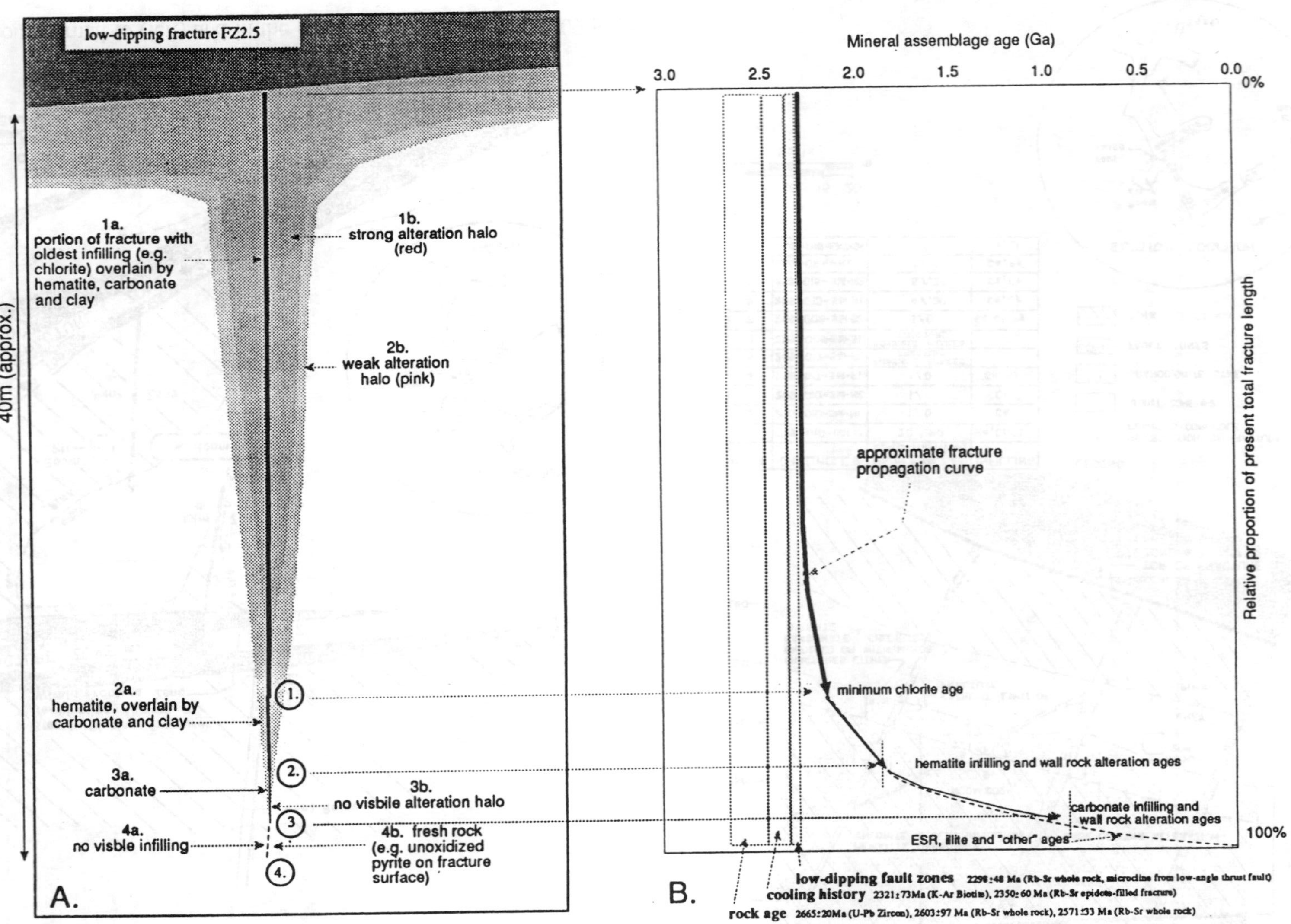

Figure 3: Simplified cross section of the Room 209 Fracture (A) showing the distribution of the infilling and wall rock alteration, and the propagation curve (B) showing the respective ages for the infillings and alterations.